The Forgotten Leg of Quality

This book emphasizes the importance of customer service in various industries and highlights its role in profitability. It presents numerous examples to remind leaders of the detrimental impact of neglecting customer service and the need for proactive measures.

The Forgotten Leg of Quality: Customer Service explores the link between customer service theory and real-world issues, such as dissatisfaction, failure, and cost. It highlights systemic failures caused by noncompliance with regulations and standards and provides strategies for implementing them. Methodologies and tools are provided to help improve recall rates and minimize nonconformances.

Through an analysis of customer satisfaction, this book utilizes various statistical studies and methodologies to aid customer service practitioners in enhancing their reporting and evaluation processes.

Practical Quality of the Future: What It Takes to Be Best in Class (BIC)

Series Editor: D. H. Stamatis, President of Contemporary Consultants, MI, USA

Quality Assurance: Applying Methodologies for Launching New Products, Services, and Customer Satisfaction
D. H. Stamatis

Advanced Product Quality Planning: The Road to Success
D. H. Stamatis

Automotive Audits: Principles and Practices
D. H. Stamatis

Automotive Process Audits: Preparations and Tools
D. H. Stamatis

The Forgotten Leg of Quality: Customer Service
D. H. Stamatis

For more information about this series, please visit: https://www.routledge.com/Practical-Quality-of-the-Future/book-series/PRAQUALFUT

The Forgotten Leg of Quality

Customer Service

D. H. Stamatis

CRC Press
Taylor & Francis Group
Boca Raton London New York

CRC Press is an imprint of the
Taylor & Francis Group, an **informa** business

Designed cover image: Pexels

First edition published 2026
by CRC Press
2385 NW Executive Center Drive, Suite 320, Boca Raton FL 33431

and by CRC Press
4 Park Square, Milton Park, Abingdon, Oxon, OX14 4RN

CRC Press is an imprint of Taylor & Francis Group, LLC

ISBN: 978-1-041-09539-2 (hbk)
ISBN: 978-1-041-09531-6 (pbk)
ISBN: 978-1-003-65061-4 (ebk)

DOI: 10.1201/9781003650614

Typeset in Times
by KnowledgeWorks Global Ltd.

To "K"

and

Timothy

Contents

List of Figures

List of Tables

Author

D. H. Stamatis, PhD, has been an academic teacher/professor at several universities in the United States, Europe, and Asia. He taught various topics, including problem solving, team approaches, SPC, G8D, 6σ, statistics, reliability, quality, and several approaches to effectiveness in manufacturing/defense/service/health and educational organizations. Dr. Stamatis has written numerous articles and over 60 books dealing with general and specific issues regarding quality. He has participated in many domestic and international quality conferences. In his consulting business, for more than 40 years, he has visited more than 120 countries, teaching, lecturing, and consulting with organizations on the benefits of focusing on quality for successful results in both productivity and profitability.

Acknowledgments

There are many individuals who helped me write the book in both encouragement and critiquing the earlier drafts. I owe all of them much gratitude and a thank you.

Special thanks, however, go to Dr. R. Munro for his never-ending discussions about quality, and certainly about audits and standards.

Thanks also go to Mr. Prat for his suggestions on the coverage of CS statistics and overall flow of the material.

Thanks to CRC Press for allowing me to use some previously published material from my own books on audits, OEE, and quality. In addition, I want to thank the editorial staff of CRC for the incredible job they have done in bringing the book to publication. Thanks to all of you.

Thanks to Tom Geraghty, CEO/Founder of *Psych Safety,* for giving me permission to use the simpler model of the OOAD.

Thanks to Rick for setting up the original format of the book.

Thanks also to Caitlyn and Cary for helping me with the graphs and figures. Their help was greatly appreciated,

And thanks to Carla Jeane for editing some of the previous drafts, even though she was in the hospital and rehab center for a significant time during the writing of the book. Unfortunately, she did not have a chance to see the final copy, as she passed to eternity. May her memory be eternal!

Preface

In 1996, I wrote *Total Quality Service: Principles, Practices, and Implementation*. It was received well by the Quality community as it was practical and covered key elements of the expected "service" in quality, including several specific tools in evaluating and tracking that service.

In the 28 years that have passed since the publication of the book, much has changed, and much has remained the same. It is the intent of this update to bring the focus on Customer Service (CS) and the need for the Voice of the Customer (VOC) to be heard, and recognize its value to the organization that is using it, as part of the Continual Improvement (CI) road to overall excellent quality.

The traditional customer service concept was to *take care problems of returns* in a counter at some corner in the back of the store. It has been used as a *detection methodology*. In this book, we have extended the definition to include issues of engineering and production in terms of non-conformances. The notion of the modern customer service is to learn about improvement and use information from multiple sources, including warranty, as early as possible. In other words, we approach it as a *method of prevention*. That is the reason why we have included discussions about specific improvement methodologies as well as specific tools. Hopefully, these can help the organization to improve and minimize or even remove potential problems from their products and or services and thereby delight the customer. So, in this book specifically, we view customer service and the voice of the customer as an important element from sourcing to delivery. We believe that building an excellent customer experience does not happen by accident. It happens by design, and the earlier we begin the process, the better the results.

It is very fascinating and quite interesting to observe that ALL organizations use many slogans and euphemisms in favor of customer service. Words like: Superb, 100% satisfaction, best in class, uncompromising quality, and so on. Yet, daily we hear catastrophic failures, and thousands of complaints/problems that consumers/customers are faced with, for products and services are offered by both large and small enterprises. A big one is the recalls from the automotive industry; a second one is the failures of the aerospace industry, and many more.

According to the National Highway Traffic Safety Administration, the data shows that the average number of vehicle recalls each year jumped 46% over a 10-year period ending in 2022, when compared with the average over the preceding 10 years (https://www.nhtsa.gov/sites/nhtsa.gov/files/2023-03/2022-Recalls-Annual-Report_030223-tag.pdf). To be sure, the vehicles are much more complicated and, as such, more problems are encountered. So, over the five years ending in 2022, the USA averaged more than 1,000 car recalls each year, or about 27 per day, the data shows. Until 2016, the USA had not exceeded 1,000 car recalls in any year going back to at least 2002. For the 2023 year, the results are not much better. According to Sedgwick's research team (https://www.sedgwick.com/

press-release/us-product-recalls-surge-in-2023-to-hit-seven-year-high/), some of the 2023 recalls are:

- Automotive recalls decreased for the second consecutive year in 2023, falling from 955 in 2022 to 916. However, the number of units recalled increased 23% year-over-year, to 38.43 million in 2023, led by electrical system issues.
- The number of consumer product recalls reached a 12-year high in 2023, with 322 events. Similarly, the number of units impacted reached a seven-year high at 135.2 million units recalled in 2023. It was also a record year for fines, with $55.3 million issued.
- There were 506 Food and Drug Administration (FDA) food recalls in 2023, a five-year high for the industry and 19.6% more events than were recorded in 2022. Undeclared allergens remained the leading cause of food recalls in every quarter of 2023. U.S. Department of Agriculture (USDA) recalls also increased, rising 41.6% year-over-year. The volume of units recalled also increased by a more notable 132.8% from 2022 to 2023.
- The U.S. medical device industry experienced a second consecutive year of growth in recall activity, rising to 975 events, with three out of four quarters dominated by quality-related incidents. However, the number of impacted units fell to a six-year low, with 283.44 million.
- Recall events in the pharmaceutical industry reached a 19-year high in 2023, rising 42.4% from 2022 to 2023, driven by an increase in sterility-related incidents. In contrast, the number of units impacted fell 82.6% to 98.5 million in 2023.
- According to the Bureau of Aircraft Accidents Achieves (B3A), there have been reported 1,416 deaths and 325 incidents for the 2021–2024 period. Cyberattacks and Software Failures Add to the Chaos. The result is that the consumer pays. (*Royal Examiner*. October 7, 2024. "Problems Persist in the Airline Industry: Challenges Continue to Disrupt Travel."
- Although the number one rule in retail CS is: *The customer is the king* through the recognition of: product, price, place, and promotion, the statistics show a different picture. Some of the top issues include long wait times, unresponsive agents, bad customer service, lack of self-service options, and poor product or service quality. The Zendesk Benchmark data shows that 95% share bad experiences and 87% share good experiences with others. A total of 54% shared bad experiences with more than five people. A total of 42% of Americans will stop shopping with a brand after just two bad experiences.

We can present many more examples; however, the point is that it seems that although "the Customer" is being talked about as the "king" and "driver" of everything that an organization is undertaking, the reality is that *profit*, *volume,* and *price* dictate the path of many organizations.

So, in order to internalize the concept of CS and CI, we must know and identify who and what the customer needs or wants. One way to understand is through Maslow's theory of motivation and human needs. Furthermore, to extend that understanding into the concept of "customer service," we must be cognizant of the **PACT** principle. PACT is essential in this understanding because "customer service" is very perishable and it depends so much on the moment of service. PACT is the acronym for (**P** = prompt, professional, personable); **A** = (accurate, attentive, acknowledging); **C** = (courteous, caring, concise); and **T** = (tailored, timely, and thoughtful). Maslow's theory is based on a hierarchical pyramid of needs shown as:

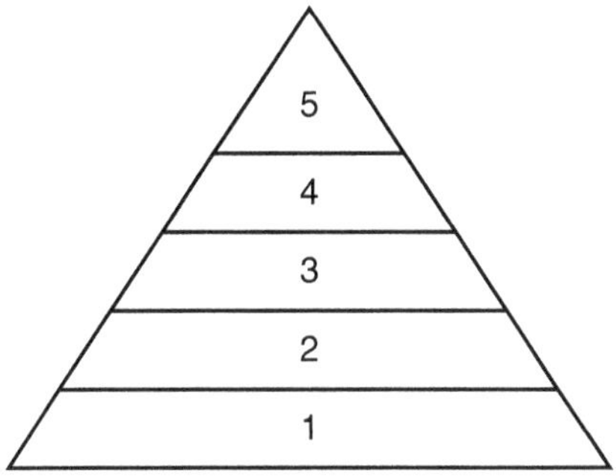

FIGURE 0.1

Levels of Needs	**Motivation and Behavior**
Level 1 = Your Body: PHYSICAL SAFETY AND SECURITY Need to stay alive! Biological and cultural imperatives to live. Includes having enough healthy food, air, and water to survive.	**Level 1** = Your Body: SURVIVAL SKILLS Eat, sleep, and take care of your bodily needs, provide for clothing, shelter, comfort, and be free from pain.
Level 2 = Your Family and Work: SOCIAL SAFETY AND SECURITY Need to be safe from physical and psychological harm in the present and future, and trust in a predictable future.	**Level 2** = Your Family and Work: SURVIVAL SKILLS Work, save for future, improve skills and talents, be responsible, and want an organized predictable world.
Level 3 = Love and Relationships: COMMUNICATION AND RESPONSE Need to be loved and to love. Includes the desire for affection and belonging.	**Level 3** = Love and Relationships: VALIDATION Join and be active in clubs and groups, be able to talk to others, contribute to society, marry, and have a family.
Level 4 = Self-Esteem: SELF-RESPECT AND ACCEPTANCE Need for reputation, prestige, and recognition from others. Contains the desire to feel important, strong, and significant.	**Level 4** = Self-Esteem: BRAINPOWER Display your talents and skills, have self-confidence, appreciate attention and recognition from others.
Level 5 = Self-Actualization: FULFILLMENT OF GOALS AND DREAMS Need for self-fulfillment. Desire to realize your full potential and become the best you are capable of becoming.	**Level 5** = Self-Actualization: CREATIVITY Be a self-starter, have enthusiasm, be creative, be dedicated, enjoy challenges, love to accomplish results!

A pictorial view of Maslow's theory may be shown in the principle of cascading need, thus:

Cascading the Y = f(X)

Level 1 $Y = f(x_1, x_2, \ldots x_n)$

↓

Level 2 $y = f(x_1, x_2, \ldots x_n)$

↓

Level 3 $y_i = f(x_i, ix_{ii}, x_n)$

Where:

Y is influenced by several potential Xs
The y is the new requirement based on X_1's and so on. The X_1 is further examined for more potential xs
The y_i is yet another requirement of the customer based on x_1's and so on. The x_i is further examined for more potential xs

Example: High-level application of a "meal in the restaurant":

Level 1 Y = a good meal in the restaurant is depended on X1 = price, X2 = service, X3 = satisfaction
Level 2 y = the result in satisfaction (which is the X3) is depended upon x1 = service and x2 = food selection
Level 3 y_i = the result in satisfaction (X3) with service (which is the x2) is depended upon x_i = availability of food, xii = management
Level 4 and so on

Another approach to defining customer requirements is the use of the Kano model. The model focuses on emotional responses that customers may care about. They are:

- Must-be (or must have) features.
- Performance features.
- Attractive features.
- Indifferent features.
- Reverse features.

The critical item that one benefits from the use of the Kano model is the realization that, over time, the delight items become basic. For example, in product development, it may be used to:

1. *Identify customer needs and wants*: Use customer feedback, surveys, and other research methods to understand customer needs and wants.

2. *Categorize features*: Map out your product features and categorize them in basic, performance, and delightful.
3. *Prioritize development efforts*: Focus on developing Basic Features first, then Performance Features, and lastly, Excitement Features. The priority should be based on some sort of information from the customer.
4. *Continuously monitor and improve*: Keep track of customer feedback and satisfaction levels to ensure that your product remains relevant and competitive.

So, by understanding Maslow's hierarchy and using the Kano Model, the organization's leadership will be able to prioritize the development efforts and improve customer satisfaction, leading to increased revenue and market share.

Everyone talks about the Voice of the Customer (VOC) and customer satisfaction. However, research indicates that these terms are acceptable, admirable, they fall into the margins of improvement for many organizations. What is needed is a true *Revolutionizing Customer Insights*. This means that the VOC must be understood and listened to. It must be a deep, beyond-the-obvious understanding of customer needs, because it is an essential component of creating competitive advantage, successful new products, and an unparalleled customer experience, in satisfying their needs and wants. In our modern world, which is a transformative period – especially with the usage of artificial intelligence (AI) – with all the advances in technology and statistical analyses, it is possible to generate this understanding.

After all, customer needs are the benefits sought by the customer when they use a product or service, interact with "a" brand, or conduct business with "a" company. Therefore, the primary goal of VOC market research is to systematically understand in-depth, beyond the obvious, important, unmet customer needs – and to let those needs drive product development and customer experience strategy. In other words, VOC allows you to focus on solving the customer problems worth solving. A very good tool for such an endeavor is the use of the *Kano* model. The *Kano* model is an analysis tool that enables you to understand how customer emotional responses to products or features can be measured and explored.

Traditionally, customer needs are best collected through surveys and or in-depth customer interviews, and sometimes with focus groups. However, many companies may be sitting on a gold mine of existing, already available, customer insights: this data comes from over 3 billion active social media users, over 100,000 online forums, and over 265 billion customer service calls placed every year for questions, comments, and concerns about products. This gold mine can be and should be taken advantage of with the technology of AI.

The average adult spends more than two hours a day on social media and has 11 customer service interactions per year (that is about one per month). Over 60 million people have posted reviews online about products and services that they use regularly. The question becomes: how can AI help with VOC insights? Researchers from Applied Marketing Science (AMS) explored and continue to explore this question in partnership with researchers from MIT and Northwestern University. This research is ongoing.

In addition to the AI, can present organizations use Machine Learning (ML) to replace traditional VOC for Consumer Products? Early-stage research indicates yes, it is possible. In their first experiment, the team tested the hypothesis that online reviews could serve as a source for consumer VOC insight. In other words, the online product reviews could produce the same results as traditional, one-on-one interviews? In this experiment, the team mined customer verbatims from over 200,000 Amazon reviews for personal oral care products. Here is an example of a typical Amazon review. The customer says, "I replaced an old brush with a new one, but the description does not say that this model no longer has a 30-second timer. The brush shuts off after two minutes, but the 30-second timer is missing. I would not have purchased this product if I had known." By looking at this review, VOC methods, applied by human analysts, could formulate a customer need, such as: "I know the right amount of time to spend on each step of my oral care routine" (https://ams-insights.com/article/revolutionizing-customer-insights-the-role-of-ai-in-accelerating-voc/).

In this process, the machine acted as a sophisticated data reduction tool, identifying the sentences in each review that were and were not informative, *e.g.*, those that contained customer needs. The machine synthesized a massive amount of data into relevant content to be reviewed by a team of research analysts.

This experiment found that online product reviews contained the same – and in some cases even more – information than researchers identified in traditional interviews. Moreover, this AI method identified customer needs quickly and effectively from an extremely large body of textual content. It is important to mention that the algorithm surfaced both frequently and infrequently mentioned needs. For example, many customers' needs centered on the desire for whiter teeth and fresher breath. However, the AI algorithm can identify lesser-mentioned needs, such as having oral care products that match one's bathroom decor. Therefore, using AI to process volumes of readily available text presents unique and exciting opportunities for those collecting insights to inform innovation and customer experience.

On the other hand, Machine Learning's Application in VOC for Business-to-Business (B2B) Markets is used extensively in many businesses where sufficient text data is available. By way of example, AMS successfully utilized its machine learning algorithm to uncover systematic and comprehensive insights in the snowplow and snow spreader category. In this case, the objective was to identify customer needs to inform product development and marketing. With the help of machine learning, the research team identified 117 distinct insights through this process. The client for this engagement was a veteran of the snowplow industry. At the onset of the engagement, he was admittedly skeptical that the methodology would yield insights that he had not heard before (https://ams-insights.com/article/revolutionizing-customer-insights-the-role-of-ai-in-accelerating-voc/).

Using machine learning combined with human analysts, AMS was able to unearth several new and exciting insights. Notably, they pinpointed a critical pain point related to visibility – the exact moment when the snowplow operator needed to have the best visibility was actually the point where they had the worst visibility. Specifically, the customer said: "The reason I don't angle it [referring to the plow] is that I hate to have one side sticking out so far, and it will block the headlights at the same angle." The customer need formulated by human analysts was: "Being able to maintain full visibility at all times, even while my plow is angled."

Other novel insights uncovered included the need for a sidewalk plow that is easy to turn around a corner; being assured the user has de-icing material for different types of roads (gravel, paving, *etc.*); and sand that will not strip away paint when applied (*e.g.*, over parking lines). In this B2B (Business to Business) example, the algorithm was able to identify new insights that were critically important to the industry (https://ams-insights.com/article/revolutionizing-customer-insights-the-role-of-ai-in-accelerating-voc/).

Introduction

Customer complaints and warranty are a fact of life in any industry or service. Even though manufacturers and service organizations would prefer not to receive complaints, they do come with a silver lining. That is: Because they are *real* customer concerns, the organization has an opportunity to improve the service and or fix it. Unfortunately, many organizations overlook both and instead of fixing the issues/concerns/problems, they ignore them, patch them and or push their resolution to some future indefinite date. Rather than managed them effectively, they ignore the potential benefits they could receive if they seriously confront them for positive results. They overlook the true role they play in the overall scheme of quality, as they are truly (or provide) a legitimate and important methodology in improving their products (services) and processes over time – that is, for companies that take a proactive stance on customer quality management, and make it a strategic function of the organization and communicated with focus on it: This strategy within the Quality domain should be directed in helping to minimize risk and reduce their program's cost, effort, and complexity.

When we consider that 20% of every dollar within the electronics industry is wasted on scrap, rework and returns, one wonders: how do you reduce these poor-quality costs and protect your bottom line? Easily, with an electronic quality management system (EQMS). You can minimize or even eliminate variations and waste during the production process by setting global quality standards through better documentation and training.

Unfortunately, only 26% of accounting professionals believe their existing compliance and controls management processes add significant value to their organization, according to a 2023 FloQast survey. If CFOs and finance executives want to get the most out of compliance and maintain audit readiness, they need agile, well-run compliance processes that allow them to do more with less. A good approach to follow is the WECARE model:

- **W**elcome: Gladly and cordially received; to greet the arrival of (a person, guests, *etc.*) with pleasure or kindly courtesy; to receive or accept with pleasure; regard as pleasant or good; to welcome a change.
- **E**mpathize: To empathize is to understand or relate to someone else's emotional experience. If you get teary-eyed upon hearing about the death of your friend's pet hermit crab, you are probably empathizing with your friend - unless you are just bored to tears by her story.
- **C**orrect: Correct, accurate, exact, precise, nice, right mean conforming to fact, standard, or truth; correct usually implies freedom from fault or error; correct answers; socially correct.
- **A**ctively Listen: Active listening requires you to listen attentively to a speaker (customer), understand what they are saying, respond, reflect on what is being said, and retain the information for later. This keeps both listener and speaker actively engaged in the conversation, and it is an essential building block of compassionate leadership.

- **R**espect: Treating someone with respect means:
  - Showing regard for their abilities and worth.
  - Valuing their feelings and their views, even if you do not necessarily agree with them.
  - Accepting them on an equal basis and giving them the same consideration, you would expect for yourself.
- **E**ncourage: To give courage, hope, or confidence to; embolden; hearten; to give support to; be favorable to; foster; help.

This model works best in problems already identified by the customer and are in the process or reconciliation (a negotiation stage). This model in particular is an excellent one to follow, for retail business where the CS department is in some corner of the store and there is a lineup of customers waiting to be served.

The results of this failure of recognition (*i.e.*, that the Customer Service) are a burden to the organization and are exemplified by the sever failures of all industries. Perhaps one of the most forgotten or rather overlooked principles of quality in the 21st century is the principle of customer service as it relates to overall quality.

Many reasons are given for failure to recognize customer service. However, in this book, we recognize that error in judgment with "minor effect" and error in judgment with "major effect." Unfortunately, many treat customer service as a catch-all problem fixing without thinking that in effect practice their own theory of GOTCHA. They fail to understand, or more likely, their failures are not accidents of performance either by machines or people, but rather purposely generated to fulfill some artificial performance without regard to Standard Operating Procedures (SOPs), regulations, and/or standards.

Do not doubt this. If we look at the statistics of failures in many industries, we see repeatedly failures that could have been avoided but did not because of poor quality control (detection) rather than quality assurance (prevention). Some examples are: Automotive, Aerospace, Pharmaceutical, Food, packaging companies, and many other sectors of our economy, daily are informing us of failures some even catastrophic. In the case of automotive, how is it possible to have so many recalls due to failures? For specific details see Visit NHTSA.gov/recalls. Let us not forget that it is very interesting that all manufacturers globally have recalls.

In the aerospace industry, how is it possible to have a door coming off the plane while flying and so many other similar instances? For detailed recalls, see Federal Aviation Administration.gov.

In the Pharma industry, we have recalls for wrong medicines and wrong applications. A drug recall is the most effective way to protect the public from a defective or potentially harmful product. A recall is a voluntary action taken by a company to remove a defective drug product from the market or warn patients and consumers about a potential risk. For specific recalls and information, see: The FDA's role in drug recalls and visit Recalls, Market Withdrawals, & Safety Alerts for all FDA-regulated products.

In the food industry, we have quite often recalls of contaminated food with listeria, salmonella, and other infectious diseases, as well as being undercooked. For detailed cases, see the FDA (Federal Food and Drug Administration).

One wonders how it is possible to have so many problems at a time when we have so many standards (International, National, Industry, and Corporate), regulations, and procedures to AVOID these catastrophic failures. We also have audits (internal, layered, third party, and so on), and still, we have problems. The problems obviously cannot be random. I submit that they are purposely overlooked in the name of productivity and profitability and lack of accountability as well as responsibility from the management team. Therefore, to truly understand the Voice of the Customer insights, an organization must have (a) management's commitment and (b) the urge to satisfy the customer by having:

1. *Compassion*: Must have or provide a heartfelt experience to every customer, member, and each other.
2. *Innovation*: Must continually pursue what is possible through research, education, operational, and service excellence.
3. *Respect*: Must honor the commitment to the communities by acting with integrity, courage, and inclusion.
4. *Results*: Must reliably deliver on the promise to be leaders in safety and affordability, ensuring the health of their communities.

So, what is customer service (CS) exactly? Customer service is the act of providing support to both prospective and existing customers. Customer service professionals commonly answer customer questions through in-person, phone, email, chat, and social media interactions and may also be responsible for creating documentation for self-service support. Many organizations can and do also create their own definitions of customer service depending on their values and the type of support they want to provide. For example, an organization may define customer service as the act of providing timely, empathetic help that keeps customers' needs at the forefront of every interaction. So, now the question is: what are the key ingredients of "excellent" customer service?

One of the many challenges in improving CS quality is to get management to view the business *at the customer level* and to develop a one-to-one mentality (attitude) toward "the" customer. *They must strive to never disappoint a single customer.* Organizational leaders who are serious about quality service take "quality" as a personal responsibility and commitment. They regard their service and their companies as an extension of themselves and their own integrity and credibility, and they spread that message throughout the organization.

Real quality service, however, is not sales slogans, buzzwords, or lip service. It is a deep, organization-wide commitment to achieving the best. It starts with a personal and professional commitment by the CEO and is ingrained in the organization's culture to the point where every employee feels a strong sense of commitment. This is particularly important for companies such as large retailers, airlines, or banks with legions of front-line workers subjected to tremendous daily pressures from customers and managers. Many companies rely most on the people who are paid the least to leave the last impression with customers. A classic example of this situation is the housekeeping staff of the hospitality industry. The least paying job controls the most fundamental quality principle, which is cleaning. The chances are slim that front-line workers will provide a satisfying experience for customers during a hassle-filled day if they do not feel there is a strong commitment to service at the top of the company. As dim as this sounds, there are some practical steps an organization or a company can take to begin and maintain the process of delivering quality service. They are:

- **Commitment**: Commit to improving quality and providing the resources to make improvement possible.
- **Delegate**: Designate a person or establish a unit reporting directly to the top. Look at everything as if you were dealing with customers or clients on a one-on-one basis.
- **Customer information**: Develop a comprehensive marketing information system through the utilization of relational database management technologies.
- **Internal assessment**: Conduct internal assessments to determine how capable your company is of delivering quality service and satisfaction. Your own research will reveal employees' perceptions of how well they are doing, organizational commitment to quality, company performance standards, how employees view customers, and what they believe their customers believe. Two excellent tools for this are: Cost of quality and the layered audit. [a. note. Cost of Quality is the sum of Cost of Good Quality (COGQ) and Cost of Poor Quality (COPQ). COGQ = Prevention Cost (PC) + Appraisal Cost (AC). On the other hand, the COPQ = Internal Failure Cost (IFC) + External Failure Cost (EFC). A Layered Process Audit (LPA) is a quality management system that is designed to add layers of protection against critical risks such as rework, customer complaints, costly product recalls, and reputational damage. It may be used in all industries and service organizations.]
- **Action**: Initiate a feedback system and use the information to adjust operations to respond to customer needs, wants, and expectations. Change procedures and even long-held policies, if necessary, and set more rigorous performance standards.

Companies that do not provide quality service and customer satisfaction, not only will not compete, but they will not exist. Just as true wisdom is knowing how much you do not know; true quality is knowing you can always do better. For quality service to exist and flourish, everyone in the organization must internalize the concept that ***quality is a journey***, not a destination, and it involves every individual in all aspects of the organization.

In the final analysis, any organization that does not take advantage of the VOC will fail in the long run. It is imperative, therefore, for the leadership team (executives) of the organization to take advantage of:

- *Appropriate tools for analyzing data*: Harnessing the most accurate and up-to-date information, businesses can identify trends and reliably forecast demand. Investing in business intelligence and analytics software offers top-tier insights into key performance metrics, ensuring well-informed decision-making and proactive risk management. Continuous monitoring of market dynamics and customer behavior is essential for both growth and profitability.

- *Embracing innovation and flexibility*: Companies must cultivate a mindset of ongoing innovation to remain relevant and competitive. Executives – especially the CEO and CFO must create plans that are flexible and can be adapted based on new insights (customer wants), achieved through tactics such as continuously testing various pricing strategies within targeted market segments. [Perhaps the most innovative act in business is the PEPSI experiment in the mid-60s. PEPSI was losing money and share of the market. The president introduced Michael Jackson (celebrities up to this point were unheard off) as the key spokesman for the company for 5 million dollars to do advertisements for a year. The *out of the Box* thinking made Pepsi phenomenal profits, and it became the #1 soft drink company in sales – at that time.]
- *Drive success with expert professionals*: Professionals bring extensive experience and specialized knowledge, enabling them to anticipate market shifts, manage risks, and identify growth opportunities. Having a strong team and infrastructure in place ensures businesses can make informed decisions and consistently stay ahead of the curve – and ultimately drive their organization to success.

Finally, for CS to work toward improvement, management must recognize that decision(s) must be taken by both management and team members. That means that freedom to choose and decide is very important, especially for the non-management person involved. For sure, there are going to be failures and wrong decisions made. However, the role of management at these crossroads must be that of encouragement, couching, and training. After all, decision making is inherently a trade-off of time and risk, as well as a balance of who is or is not involved in the decision-making process.

1 General Customer Service Overview Comments

OVERVIEW

One wonders why customer service is important – a challenging question but very essential to do business. When 86% of customers (https://www.helpscout.com/75-customer-service-facts-quotes-statistics/) quit doing business with a company due to a bad experience, it means that businesses must approach every support interaction as an opportunity to acquire, retain, or up-sell. Good customer service is a revenue generator. It gives customers a complete, cohesive experience that aligns with an organization's purpose.

According to a variety of studies, US companies lose more than $62 billion annually due to poor customer service management, and seven out of ten consumers say they have spent more money to do business with a company that delivers great service (https://www.helpscout.com/75-customer-service-facts-quotes-statistics/).

Therefore, understanding that customer service is the cornerstone of your customer experience (CX) helps you leverage it as an opportunity to *delight* customers and engage them in new, exciting ways. So, what are the principles of good customer service? From our experience, there are four key principles of good customer service: Being personalized, competent, convenient, and proactive. These factors have the biggest influence on the customer experience. Let us examine them.

- *Personalized*: Good customer service always starts with a human touch. Personalized interactions greatly improve customer service and let customers know that your company cares about them and their problems. Instead of thinking of service as a cost, consider it an opportunity to earn your customer's business all over again.
- *Competent*: Consumers have identified competency as the element that plays the biggest role in a good customer experience. To be competent, a customer support professional must have a strong knowledge of the company and its products, as well as the power (authority and responsibility/accountability) to fix the customer's problems. The more knowledge they have, the more competent they become.
- *Convenient*: Customers want to be able to get in touch with a customer service representative through whichever channel is the most convenient for them. Offer support through the channels of communication your customers rely on most, and make it easy for customers to figure out how to contact you.

DOI: 10.1201/9781003650614-1

- *Proactive*: Customers want companies to be proactive in reaching out to them. If one of your products is backordered or your website is going to experience downtime, proactively reach out to your customers and explain the problem. They may not be happy about the situation, but they will be thankful that you kept them in the loop.

By building your customer service strategy around these four main principles, you will create a positive, hassle-free customer experience for everyone who deals with your company. The concern(s) question(s) that arise dealing with CS, especially when failures and lack of customer satisfaction, fundamentally lead us to organizational leadership and disregard of regulations, standards, policies, instructions, lack of responsibility, accountability, and lack of appropriate training.

Therefore, the issue of any CS program is fundamentally a *leadership concern*. This means that the management executive team must operate under principles that promote:

1. *Encouragement* for innovation and curiosity. They must ask continuous questions as to what can be improved with new ideas and challenge the *status quo*.
2. *Communication* with empathy and transparency. Understand the thoughts and feelings of others, and be open and honest.
3. *Esurance* for outcomes and accountability. Commit to achieving specific results through collaboration and data-based decisions.
4. *Ownership* of safety and reliability. Embrace high standards to ensure excellence in care and service.
5. *Authenticity* with both behavior from both the head and heart. Act with recognizing that "a" risk vulnerability exists in the decision(s).
6. *Coaching and inspiration* to do the best. Encourage all employees to grow and develop. Practice productive feedback.
7. *Integrity and trust* by honoring commitments and confidentiality (when appropriate) and always deliver on promises.

Furthermore, to start focusing on customer satisfaction, the organizational leadership must ask three very fundamental questions that demand honest answers. These are:

- **Why**: Why are we doing a particular task? If we do not understand the "why we are doing something," we are doomed to fail.
- **How**: How are we doing a particular task? Can it be done differently? More efficient? What is stopping us from changing?
- **What**: What is the result of what we are doing now? Are we satisfied? Is the customer satisfied? If not, why not? What can we do as an organization to make the customer satisfied, happy, maybe even delighted with our product(s)?

So, now comes the self-evaluation of the three basic questions: (a) How does your organization articulate performance standards? (b) Are they doable/practical?

Or are they theoretical with some abstract deadline and goal, in your organization? and (c) Is it a long list of rules, standards, metrics, and infractions or goals? Or are they concrete, defined, and above all realistic (doable), for both your organization and your customer(s)?

However, to do this the self-evaluation, one must have a strong personality with flexibility and a realization of being wrong, sometimes. Therefore, to be successful, the individual must have the following characteristics for his/her own psychological safety. The individual leader unquestionably must operate from a level that creates, facilitates, establishes, and embraces change in the organization. Specifically, it means that is open to discussion(s) regarding quality and specifically customer service satisfaction through the following personality traits:

- Feel free to express his/her opinion(s)
- Feel safe of negative responses
- Feel confident of her/his mistakes, recognize them, admit them, and if possible, correct them
- Feel safe to take risk(s)
- Encourage others to be honest, open-minded, and freely discuss their opinion(s)

In the manufacturing world to ensure quality, there are regulations, standards, and audits that "supposedly" control the overall quality and monitor the satisfaction of the customer. Four of the primary methods are to use: (a) the PPAP (production part approval process) approach, (b) the APQP (advanced product quality planning) methodology, (c) MSA (measurement system analysis, and (d) the OEE (overall equipment effectiveness) process for machine capability and efficiency. On the other hand, in service organizations typical methods and tools are the seven basic tools of quality and the 8D methodology. In Chapters 8–10, we discuss these and some other methodologies/tools (however, not an exhaustive list) in evaluating CS as well as correcting and/or improving deficiencies in the organization's system.

In the dynamic landscape of the life sciences industry, ensuring compliance with good manufacturing practices (GMP) is imperative to guarantee the safety, efficacy, and quality of pharmaceutical products. One critical aspect of GMP is the analytical product quality review (APQR), an essential process for evaluating and monitoring the consistency of product quality throughout its life cycle. APQR is also known as annual product review (APR) in the U.S. Food and Drug Administration, and product quality review (PQR) in EU good manufacturing practice (GMP) guidelines. It is required to be conducted annually for every product to determine the need for changes in drug product specifications or manufacturing or control procedures.

However, challenges in the industry related to PQR have highlighted the need for innovative solutions. These are a few of the current challenges while conducting effective APQR.

1. *Manual processes and data management*: Traditional PQR processes often rely on manual data collection and analysis, leading to inefficiencies, errors, and delays. The sheer volume of data generated in the manufacturing

process, such as data from quality control, regulatory compliance, equipment, and environmental monitoring, makes it challenging to manage and analyze information accurately.

2. *Diversity in templates across sites and product types*: The use of multiple templates across different sites and product types introduces complexities in data standardization and analysis. This diversity hampers the ability to derive meaningful, standardized insights from APQR reports.
3. *Annual document preparation and delayed detection*: Document preparation occurring annually, and the absence of real-time, actionable insights, hinder the industry's ability to respond promptly to emerging quality issues, potentially affecting product quality and patient safety. The delay in detecting and correcting trends or shifts also results in costly recalls and regulatory actions, posing a significant risk to product quality and compliance.
4. *Database maintenance challenges*: Managing a vast database with one document per year for each product becomes a significant challenge. This approach requires extensive storage facilities, which adds real estate cost and can impede efficient retrieval and analysis of historical data because there is a regulatory requirement of maintaining the approved APR documents for a minimum of 11 years.
5. *Integration hurdles for electronic adoption*: The transition to electronic APQR is hindered by integration challenges, making it difficult for companies to seamlessly incorporate electronic processes into their existing workflows.
6. *Regulatory compliance*: Meeting the stringent regulatory requirements for PQR is a complex task. Frequent changes in regulations along with the need for comprehensive documentation add a layer of complexity to ensure compliance with various global regulatory bodies.

THE VERY BASIC APPROACH TO CS

We all know mistakes happen. However, how we recover is very important for the organization. We think that the principles for appeasing, and ultimately having a happy customer, are the universal personal approach of kindness, listening, and temperance/attitude of your behavior to the specific situation. That means take personal responsibility for your attitude and behavior. We propose the following introspective actions as appropriate considerations to start with:

1. I am positive; *i.e.*, I must display a positive attitude toward an angry and dissatisfied customer.
2. I am accountable for success; *i.e.*, resolving the issue is my responsibility. I will do what it takes.
3. I am a good communicator; *i.e.*, I am willing to go the *extra mile* to resolve the issue – even if I must go to my superiors.
4. I am responsive; *i.e.*, I am listening to your concern, and I feel your frustration with the failure of the product or service.
5. I am proud to serve you; *i.e.*, *demonstrate* your proudness with words and actions to the concern(s) of the customer. [Just remember that the customer

is just as frustrated with the process as much as you. Therefore, hold your temper and be patient. Listening is the trick of good service. Demonstrate that the system you work under is a good one as well as customer-friendly system.]

6. I am respectful; *i.e.*, the customer is angry and quite often obnoxious. Therefore, it is up to you to keep your cool and act professionally. Assure the customer that all that takes place is private and confidential. At this moment, you represent the organization, and future sales depend on how you handle the interaction. As we can see the statistics, later, word of mouth travels fast. And if it is a negative experience, it will travel much faster and more likely it will be embellished. So, be patient and respectful. Listen, listen, and then listen again!
7. I am committed; *i.e.*, again, through your body language as well as the tone of your voice, show a legitimate concern (be empathetic) to the customer. Show that you want to satisfy the complaint, and more importantly, thank him/her for their business. Your enthusiasm should be obvious. Treat the customer with a smile and a thank you.
8. I am meticulous; *i.e.*, talk is cheap. Therefore, make sure that your work environment is clean and in order. Above all, make sure that all the common tools you need for a good conclusion to the problem at hand are available and working before you start the process of resolution. You will lose credibility if you are not ready.
9. I am innovative; *i.e.*, demonstrate that you can take responsibility of your actions and you do not mind to tackle problems with "out of the box" traditional approaches. Always remember it is much easier to say "I am sorry" or "forgive me" than to say "May I do this, or this…"

MODERN SOLUTIONS FOR EXPEDITING AND MONITORING THE PROCESSES

We recognize that monitoring individual processes for the complete and correct PPAP, APQP, OEE, and MSA is a complex activity. However, technology has helped tremendously in this area, and we have many software packages to tackle the tasks. Here we identify – what we think – are some of the better ones. There are many others, and we encourage the reader to investigate their availability, cost, and utility to their individual scope of operations.

For the APQP, the DELMIAWorks software is a good one. For more information, see their website: https://www.solidworks.com/product/delmiaworks/manufacturing-erp/quality/apqp/. Its features are:

- This software includes direct links to FMEAs, Control Plans, inventory items, customer information, and more. It is a centralized, integrated location to perform all the phases of APQP, including:
  - Planning & defining the program
  - Product Design and Development

  - Process Design and Development
  - Product & Process Validation
  - Feedback, Assessment & Corrective Action
  - Control Plan
- In addition, the DELMIAWorks APQP system supports the timely completion of required tasks, as well as effective communication with employees, customers, and suppliers. [Note: We make the distinction between supplier and vendor because there is a big difference between them. A *vendor* is one who does business with you on an irregular basis, and usually, the price is the driving force. On the other hand, a *supplier* is one who works with you and develops a relationship with you for the benefit of both. Price may still be important but the relationship advantage overpowers it.].

For the PPAP, a good software is the SAFETYCULTURE. For more information see their website: https://safetyculture.com/app/ppap-software/. Its features are:

- Uses PPAP checklists and inspection templates for auditing production processes and submitting production part documents for approval.
- Digitizes the document management system via cloud-based solutions to boost efficiency and save on space.
- Communicates production updates with the entire team through Heads Up to ensure the organization is on the same page.
- Gathers data and analytics on the PPAP process to find potential areas of improvement.
- Checks the quality of product parts, spots non-conformities, and assigns the right people to address them.

For the OEE, a good software is the SAFETYCULTURE. For more information, see their website: https://www.oeesystems.com/solution/oee-software/. Its features are:

- Provides a continuous improvement process that can be used on single and multiple production lines and from one to several plants in a group-wide structure. It is suitable for all types of manufacturing processes, including discrete, continuous, or bulk. It is perhaps one of the better OEE software that measures the gap between the actual performance and the potential performance of a manufacturing unit.
- Offers complete visibility of your production processes.
- Monitors and improves your OEE with data-driven insights that identify, highlight, and communicate performance issues in real-time.
- Tracks performance from the start of each shift and uses customizable reporting with interactive charts to drive continuous improvement.

For the MSA, generally, one may use the *Minitab* software tool, which is used for data analysis, statistical, and process improvement by organizations worldwide to improve quality and reduce costs. Another option is the MSA software. For more information, see website: https://www.softexpert.com/produto/msa-measurement-systems-analysis/. Its features are:

- Provides a guide to assist in the formal development of a measurement system.
- Establishes manufacturing process acceptability.
- Classifies equipment based on a range of criteria such as lab applications, place of utilization, equipment type, *etc*.
- Demonstrates how to find the root cause of measurement processes and explore causes based on investigation.
- Standardizes measurement processes involved with equipment calibrations.
- Provides standardized methods of data measurement collection.
- Maintains documented procedures and instructions for the execution of studies.
- Defines specific ranges for each type of equipment and analyzes their characteristics.
- Integrates documents, calendars, stability studies, bias, linearity, R&R, ANOVA, and other.
- Compares equipment performance before and after repairs and maintenance.
- Provides complete analysis using graphs, amplitude, errors, histograms, linearity, and others.

For the life sciences, the AmpleLogic software from Digital Solutions is a good one. Currently is (considered to be) a leading software in providing innovative solutions for the life sciences industry. It offers, in our humble opinion, a comprehensive platform designed to address the challenges associated with PQR and elevate APQR processes. For more info on this software, see website: https://amplelogic.com/product-quality-review-software/. Its predominant features are:

- *Automated data collection and analysis:* A digital APQR platform automates the data collection process, reducing manual errors and streamlining the analysis of vast datasets. This ensures the timely generation of APQR reports, enhancing overall operational efficiency. The process for such efficiency is twofold, expressed as a two-stage evaluation. The first stage is: Report generation, review of data, quality data, process data, and generation of a master data plan. The second stage is: Data extraction, which includes master data, process data (quality and manufacturing data), quality data, reviewed data (statistical analysis), and report generation.
- *Integration capabilities*: The platform facilitates the integration of diverse data sources, breaking down silos between manufacturing, quality assurance (control), and supply chain data. This integration enhances the ability to derive holistic insights, enabling informed decision-making. The ultimate idea here is to generate fast and accurate CS reports through statistical data analysis.
- *Cross-functional department collaboration*: Users can see what they have been assigned, and various department users can contribute and collaborate

on PQR. Users can send all approvals at once, receive email notifications and sign them, add notes or comments, and respond and make changes based on those notes.

- *Real-time data availability before batch release*: A digital APQR ensures that all relevant data are available in real time before batch release, enabling proactive decision-making and minimizing the risk of quality issues escaping detection.
- *Continued process verification (CPV)*: APQR facilitates CPV by providing continuous monitoring of critical parameters throughout the manufacturing process. This proactive approach ensures ongoing compliance and quality assurance.
  - Control charts are used to track critical quality attributes (CQAs) and critical process parameters (CPPs) over time. Any trends, shifts, or out-of-control points on the chart can signal potential issues that need investigation.
  - Specification limits are compared to the control chart data to ensure that the process is consistently producing results within the defined specifications. [a. note. WARNING! Specification limits should never be plotted on the control chart. The reason is that specifications are customer-given, whereas the control limits are generated from the process (they are calculated). If the specification values are plotted on the control graph, for sure, they will create confusion. They are not the same.]
  - Control limits on control charts help identify statistically significant variation in the process. Points beyond these limits may indicate special cause variation that requires investigation.
  - Although CPV includes ongoing monitoring of the process, the initial validation batch data serve as a baseline for expected performance. Deviations from the validated state may trigger further investigation and corrective actions.
  - In CPV, Ppk (process performance index) calculations help quantify how well the process is meeting specifications. A high Ppk value indicates a capable process, while a low value may suggest a need for process improvement.
- *Compliance to at least three sigma*: APQR aligns with statistical process control methodologies, ensuring compliance with three sigma acceptance criteria. This data-driven approach enhances the precision and reliability of product quality assessments.
- *Centralized monitoring of critical control points*: The APQR system establishes tolerances for critical control points (CCP) and escalates any excursions beyond proven acceptable range (PAR) or outside the normal operating range (NOR). This proactive alerting mechanism ensures swift corrective actions for every batch manufactured.
  - PAR is the demonstrated range of process parameters that consistently produces a product meeting its predefined quality attributes. It is determined through the validation process and represents the acceptable variation in the parameters that ensure product quality.

    - NOR is the specified range within which process parameters are expected to operate under routine manufacturing conditions. It is often set within the wider PAR and represents the typical operating conditions for day-to-day production.
- *Employee performance capture*: APQR goes beyond product-centric metrics by capturing employee performance data. This holistic approach allows organizations to assess and improve overall operational efficiency.
- *Golden batch facilitation*: APQR supports the concept of the "golden batch," allowing organizations to identify and replicate optimal manufacturing conditions consistently.
- *Anytime reviews and report creation:* By simplifying and automating the process, users can create virtual or interim PQRs, allowing them to routinely assess the state of their processes and review the progress of their continuous improvement initiatives.
- *Regulatory compliance management*: New-age solutions are equipped with built-in compliance features such as e-signatures and audit trails as required by 21 CFR Part 11 and EU Annex 11, ensuring that PQR processes adhere to the latest regulatory requirements. This capability enables life sciences companies to stay ahead of regulatory changes and maintain compliance seamlessly.
- *Integrated analytical functionality*: APQR integrates comprehensive analytical functionalities, eliminating the need for external analysis tools, ensuring a seamless and efficient APQR experience.

In conclusion, addressing PQR challenges in the life sciences industry requires innovative solutions to maintain product quality and GMP compliance. By leveraging cutting-edge digital technology for end-to-end automation, drug manufacturers can streamline data collection, analysis, collaboration, and approval processes.

AmpleLogic's platform stands out as a robust solution, offering automated processes, regulatory compliance management, integration capabilities, and real-time monitoring at reduced costs. Through this technology, companies can enhance their APQR processes, bridge existing gaps, and confidently navigate the pharmaceutical industry's complexities. Modernizing the APQR process also brings benefits like electronic collaboration, streamlined regulatory report compilation, time savings, and improved data analysis, which facilitates quicker and more routine informed decisions regarding product quality.

It does not matter how great your product is: If your customer service is poor, people will complain about it, and you will lose customers. The good news: It is not impossible to turn things around. Transforming your customer service from mediocre to great will not happen overnight, though. It requires a serious commitment to meaningful change, a team of rockstar support professionals, and work across the entire organization.

CUSTOMER SERVICE SKILLS

Perhaps one of the most critical skills in dealing with customer service and satisfaction is the selection of the team. That is because delivering consistently good customer service requires work and alignment across your entire organization. Please notice that there is no "I" in the team. In fact, team, if used as an acronym, means: **T**ogether, **E**veryone, **A**ccomplishes, **M**ore. That is why it is imperative to hire people who genuinely want to help your customers succeed – and to pay rates that are attractive to skilled professionals. Of course, finding the perfect hire for a support team can be challenging. No particular checklist of job experiences and college diplomas adds up to the perfect candidate. Instead, you are looking for qualities that can not necessarily be taught. These individuals thrive on one-on-one interactions within their community. They love problem solving. They are warm, approachable, and great at teaching other people how things work. For a good list, see the website: https://www.helpscout.com/helpu/customer-support-department/.

To be sure, the competition for all industries is very fierce, and that is why it is imperative to be prepared to respond to your customers with an unwavering desire to satisfy them. In other words, provide a total environment for a "great" customer experience (CX). How do you do that? Training your customer service representatives on what to say and how to say it is the first step. The truth is, your customer service representatives are the front line of your business. These are the people who represent your business and engage with your customers (potential and longtime buyers alike) day in and day out. Some deal with complains about your problems, and some deal with potential sales and problems. To address them, we suggest some key (of the many) skills available that are necessary for success. The basis for these is based on https://broadly.com/blog/customer-service-skills-employee-needs/#:~:text=Sometimes%2C%20patience%20just%20means%20listening,they%20are%20confused%20or%20frustrated, and https://www.investopedia.com/ask/answers/032615/what-are-four-types-economic-utility.asp, written by Help Scout (June 27, 2023). They are:

- *Empathy*: Empathy is the ability to understand and share in the feelings of somebody else. This is a good example of a customer service skill that tends to be more of a personality trait, but with careful attention, you can still work to improve it. This skill comes in handy because, even if you cannot give a customer exactly what they want, you can understand their frustrations. It may not solve the problem at hand immediately, but acknowledging the feelings the customer may be experiencing will help you gain the customer's trust. They will see you are a human and not just a robot with zero regard for their well-being. As soon as your employees understand how to empathize with the customer, they can work toward creating more strategic messaging to help create a better outcome for everyone involved. Something as simple as the following statement can work wonders: "I understand you're probably very disappointed that we could not honor the expired sale price, but I want to help make things right."
- *Patience*: Truly being patient with a customer means to wait without getting annoyed or upset. The need for patience at work can arise in a variety of different scenarios, no matter the managerial level of the employee. Sometimes, patience just means listening to the customer air their grievances. Letting your client feel heard and respected is a big customer service tactic that

can pay off well – especially when you consider the fact that customers are often reaching out because they are confused or frustrated. Simply take the time to hear them out without cutting them off or insinuating they are wrong, and see what happens. Always remember the old saying: *Patience is a virtue.* It can even diffuse anger, if used properly.

- *Positive language*: Nobody likes a Debbie Downer at parties, and your customers will certainly not appreciate one when they are trying to do business with you. [a. note: Debbie Downer is a fictional Saturday Night Live character who is constantly making others feel bad or dampening the mood of a group with negative comments. In fact, one may characterize her as a negative or pessimistic person. From a CS perspective, a Debbie Downer euphemism is a name for a person who speaks only of the bad or depressing aspects of something and lessens the enthusiasm or pleasure of others.] Words have power, consequences, and it is in your best interest to use your words to strengthen customer relationships, not tear them down. Using positive language in customer service means you can turn any situation into an open-ended, positive one instead of a closed-off, negative one. For example, if a customer asks you a question, and you are not sure of the answer, what do you think is the best reply? Two answers that may defuse the critical situation are:
  - I do not know
  - Great question, let me "look into that" for you! Or let me investigate this to find out what is happening here. This answer is an example of using positive language. Do you see how the second reply can turn the situation around, while the first reply is a dead-end, an unenthusiastic answer? Another example is, instead of saying you cannot do something, say that you can try to solve the issue. Again, it is a simple switching of your language, but it makes a very big difference in your customer's happiness. [A WARNING is necessary here. Everyone makes mistakes. When you make one and say, "I am sorry. That is our mistake," it is good customer service, but not service recovery. Most organizations never admit mistakes, and fewer than 1% practice service recovery – fix the problem at the source – if possible, and make the customer "happy." It is an understatement to say how important keeping customers happy, calm, and reassured that a resolution will take place. It is indeed very critical to your growth. Gaining a new customer is expensive. When you lose one, it is super-expensive. Not many employees know the lifetime value of a customer, so if one has a problem with a $25 purchase, they will shrug and say, "Big deal. It is only $25." However, the lifetime value of that customer could be $1,200, $3,500, or even $11,000. For example, the average Ritz-Carlton customer will spend $250,000 with the Ritz over their lifetime. For more information on this recovery, see: J. Tschohl (2014). *Achieving Excellence Through Customer Service.* 10th ed. Bestsellers Publishing. St. Augustine, FL; See also: https://www.qualitydigest.com/inside/customer-care-article/service-recovery-60-seconds-or-less-073124].

To employees, it may be insignificant if the customer never comes back. But the cost to the organization is huge. For calculating this cost, see Tschohl (2014).

Employees often feel that the customer is their enemy, especially when something gets screwed up. The customer gets excited and often raises their voice or even uses bad language. Is your employee trained to flip that situation around in seconds so they keep the customer? Too often, the employee does not want to lose face and admit the problem or take responsibility, or they do not know how to handle the situation. It is very frustrating for the customer when the front-line representative has no clue what to do, calls for another, they do not know what to do, they call a supervisor who also does not know what to do, and now the problem is escalated to a manager level. Had the representative been given the authority and responsibility to settle problems with a limit of say $1,000 worth with no questions asked, everything would be fine, and the customer would be surprisingly satisfied. Of course, the limit is increased for the supervisor and the manager.

So now the question is: How do we make a bad situation into a positive experience? The answer is simple – remember the age-old saying. If you are served with lemons, you can make lemonade out of the sourness of the lemons. How? By following the following steps:

1. The resolution must (if possible) be implemented on the front line.
2. The action must be fast. The sooner the problem is resolved, the greater the satisfaction.
3. The representative must be friendly but resolute.
4. The representative must have the authority and responsibility to resolve issues within a range of monetary value or exchange of the same or similar item. This is where you have a chance to delight the customer with an unsuspected surprise. Your chance as a representative to make sure the customer does not leave the counter with "I'll never use your firm again," instead, you want them to post on social media about what just happened and how awesome you are!

   [A real example from a major retailer (Underfit, Inc.) may help here: An angry customer returned a purchased item (which took two weeks to be delivered). The item was the wrong color and size. Needless to say, the customer was loud, and one may say, she was obnoxious. The CS representative was very attentive and calm, listening to the concern. When the customer was done venting, the representative, in a very calm and apologetic voice, recognized the problem and assured the customer that the item in question will be replaced the next day for no extra charges. The customer left with a smile on her face and was very satisfied. In fact, she told me that she was very impressed with the entire experience and would not hesitate to do business again with Underfit, Inc].

- *Clear communication*: Clear communication is achieved when the exact message you are trying to send is the exact message received. Sometimes, people can misinterpret confusing language and take it to mean something completely different than intended – that is the opposite of clear communication. Do not use academic, and or prosecutorial language – ever. **K**eep **I**t **S**imple, **S**tephany (this is known as the *KISS* principle). When it comes to important details, make sure you leave nothing to doubt. If you need to take extra steps or repeat something to ensure your customer understood you clearly, then go ahead and provide some extra explanation. Nobody likes feeling like they have been tricked (even if by accident).

- *Knowledge of product*: If you do not know your product, how do you communicate the benefits to your customer and get them to buy it? Obviously, it is a big annoyance for customers when a company's staff seems unknowledgeable. You should know what you are selling as if it were the back of your hand. Lack of knowledge about your product can immediately crush your customer's trust that you know what you are doing at all. Always be honest with the customer! However, if someone asks a question about your product you are unsure of, use positive language to help you figure it out: A good response may be something like: "To be honest, I'm not sure if we carry that model anymore, but let me check with inventory to make sure."
- *Able to read between the lines*: These days, customer service can be conducted via telephone or computer, which means it is much harder to read customer cues. When you cannot see facial expressions or hear the tone in a customer's voice, it is much tougher to assess their true feelings about the situation. That is why it is more important than ever to anticipate and read between the lines. Reading between the lines means thinking beyond what the customer is telling you. Look for subtle clues that may be able to help you understand their mood, patience level, or personality.
- *Persuasiveness*: Being persuasive means you can help convince the customer to take an action you want them to take. Persuasiveness does not equal being pushy. Many times, customers will fill out your customer feedback forms with curiosity and questions about your company, not because they want to complain. If your customer service representative or employee receives one of these curious inquiries, s/he needs to be ready to convince the customer that the company is the right choice. A good customer service representative should always be ready to turn a question into a sale (or complete a positive resolution of an existing problem), and having persuasion skills is the best and appropriate way to do this.
- *Calming presence*: To have a calming presence, take special note of the social cues you are exuding. Harsh tone and fast speech are two no's if you are trying to calm the customer down. "Keep it cool" under pressure, and see the positive results that ensue. If a customer is upset, the customer service support team is that customer's one hope that their issue can be solved. Instead of blaming the customer or getting testy with them, stay calm and walk them through their issue as slowly as needed. Hopefully, by the end of the conversation, your customer will also be much calmer. This characteristic is of special importance if the customer is returning something – for any reason. Without a game plan for the way your team operates and communicates with customers, your business can suffer big-time.

Every employee in your organization should exhibit these customer service skills if you truly want your company to succeed. Not only are these qualities helpful in business, but they will likely propel you and your employees in life overall as well. It is of paramount importance to recognize that these skills are not working in a vacuum or a silo. Rather, they are always working in tandem with each other. For example: It is hard to be patient without empathy, and it is difficult to retain a calming presence without patience. That means, though, as you master each one, the others will become easier. It is also important to recognize that some skills may be inherent or innate to a particular individual, and some are not. However, rest assured that all can be taught over time, and the opportunity to learn from the leadership of the organization. It just takes commitment and time, and the results eventually will surface in the financial records.

A short reminder about teams. Always remember the saying of P.T. Barnum: *A cent will make a difference for a sale.* In our modern world, through neuropsychology, we know that indeed, one cent, makes a difference. See the science behind it at: https://www.newneuromarketing.com/99-or-00-when-charm-pricing-backfires#:~:text=. Also remember the "baker's dozen" concept. A baker is practicing the notion that "if you give people more than what they expect to get," your customer will be happy and satisfied. In the Bakers dozen, the baker by placing an extra donut in the bag, the customer would receive 13 donuts instead of the full dozen of 12.

The tide is shifting in business. Many companies realize that their best employees could be poached and go elsewhere for more pay and perhaps a better employee experience. Therefore, it is essential to prioritize your customer service team members' needs and address their priorities.

Below are a few customer service statistics that emphasize just how important the level of care that you provide is:

- The customer service industry has the highest employee turnover rate of all businesses. Call center statistics have placed it as high as 45%! (https://qatc.org/winter-2015-connection/exploring-call-center-turnover-numbers/).
- The Human Resource Institute estimates that turnover costs about $10k–$15k for a frontline employee (https://qatc.org/winter-2015-connection/exploring-call-center-turnover-numbers/).
- Most customer service reps do not plan on staying in that job forever, as there are often poor growth opportunities. A total of 31% of consumers consider a knowledgeable agent to be the most crucial factor for a positive customerexperience(https://www.thomasnet.com/insights/customer-service-is-manufacturing-s-silver-bullet/ Thomasnet).
- Since 77% of consumers consider the difficulty of reaching an agent to be the biggest issue with customer service today, it is worth investing in resolving it (https://review42.com/resources/customer-experience-statistics/).
- A total of 89% of companies with "significantly above average" customer experiences perform better financially than their competitors (https://www.qualtrics.com/m/assets/wp-content/uploads/2020/04/XMI_GlobalStateOfXM-2020.pdf).

- Great client support yields long-lasting results, too. A total of 24% of consumers keep doing business with vendors for two or more years after a positive encounter (https://d16cvnquvjw7pr.cloudfront.net/resources/whitepapers/Zendesk_WP_Customer_Service_and_Business_Results.pdf).
- A total of 64% of U.S. consumers contacted customer service in 2017 (https://www.statista.com/statistics/815526/customers-who-have-contacted-customer-service-in-the-past-month-us/).
- A total of 73% of customers say they stay loyal to brands because of friendly customer service agents (https://blog.hubspot.com/service/importance-customer-service).
- A total of 73% of customers fall in love with a brand because of friendly customer service representatives (https://www.slideshare.net/slideshow/2011-customer-experience-impact-report/10947124#5).

Consider those statistics, and imagine that you have six employees at the start of the year. By the end of the year, two or three of them will likely have left. That is between $30k and $45k in expenses for losing employees who were likely doing an excellent job. As we know, good customer service representatives can make or break customer relationships, so prioritizing the people who do good work is extremely important.

Consider what your customer service team needs more of, and try to find ways to offer it. Then speak directly to your team members about changes that you have made and the impact it has had on their quality of life. Ask them for their opinions, listen carefully, and then implement what you can. Making a better working environment for your agents will create an excellent environment for your customers.

ARE YOU A TRANSFORMATIONAL LEADER?

The current economic landscape is complex. It takes strategic leaders to navigate it. That means, transformational leaders are key to driving business success. The dynamic nature of modern business requires finance leaders to serve as forward-thinking business partners who provide real-time data and insights to impact decision-making, enhance operational efficiencies, and drive value. So, to position your organization for success, you will need to prioritize:

1. Leveraging data analytics for informed decision-making.
2. Effective working capital management.
3. Investing in growth opportunities.
4. Maintaining readiness to access markets.
5. Embracing flexibility and innovation.

With an evolving economic climate, having an experienced partner can make all the difference. Quite often, hiring an individual or a consulting organization with a proven track record may be able to guide your organization through complex financial challenges, CS strategy, and certainly offering valuable insight and transparency for growth and profitability. In any case, ALL organizations should learn about strategies for organizational resilience. For example, at least the following should be included.

- *Problem-solving skills*: Customers do not always self-diagnose their issues correctly. Often, it is up to the support representative to take the initiative to reproduce the trouble at hand before navigating a solution. That means they need to intuit not just what went wrong, but also what action the customer was ultimately after. An example may help here: If somebody writes in because they are having trouble resetting their password, that is ultimately because they want to log into their account. A good customer service interaction will anticipate that need and might even go the extra mile to manually perform the reset and provide new login details, all while educating the customer on how they can do it for themselves in the future. In other situations, a problem-solving approach may simply understand how to offer preemptive advice or a solution that the customer does not even realize is an option. This is where a CAPA methodology (Corrective Action, Preventive Action) may help. Typical tools, and there are many, are:
  - PDC(S)A (Plan – Do – Check (Study) – Act).
  - SPC (Statistical Process Control).
  - FMEA (Failure Mode and Effect analysis).
  - Six Sigma Methodology (DMAIC (Define, Measure, Analyze, Improve, Control and or DMADV (define, measure, analyze, design, and verify).
  - Basic statistical approaches (check list, Frequency, Average, Pareto, Pie Chart, *etc.*).
  - Advanced statistical approaches (Correlation, regression, ANOVA, Tolerance Design, *etc.*).
  - Artificial Intelligence (AI). Whether we like it or not, AI is moving fast in all phases of our lives. CS is not excluded. With the recent popularity of large language learning models (LLMs) like ChatGPT, the Asset Management Services (AMS) platforms research team, in partnership with academics at Northwestern and MIT, is now investigating whether LLMs can formulate abstract, conceptual customer needs for customer verbatims. Previously, this was a task that only human analysts have been able to accomplish. Preliminary results from this new research are promising. We see early evidence that LLMs can formulate customer needs, indistinguishable from human-formulated needs. Notably, these needs are sufficiently detailed for product development. This specificity is critical for innovation. By harnessing the power of AI and LLMs, researchers can quickly and effectively synthesize massive amounts of qualitative data to unearth the nuanced customer needs necessary for effective innovation. While machine learning algorithms will never completely replace insights and product development professionals, they can certainly improve efficiency and effectiveness, allowing the team to focus on the most value-added portions of the insights and product development process. All of us should be excited to be so instrumental in these exciting changes that impact the way we collect, analyze, and

understand the Voice of the Customer insights. Certainly, our task is a large and significant one; however, to be successful, we must display at least the following characteristics:

1. *Compassion*: We provide a heartfelt experience to every patient, member, and each other.
2. *Innovation*: We continually pursue what is possible through research, education, clinical, and operational excellence.
3. *Respect*: We honor the commitment to our communities by acting with integrity, courage, and inclusion.
4. *Results*: We reliably deliver on our promise to be leaders in safety and affordability, ensuring the health of our communities.

- *Greatly affect how the customer hears the response*: Patience is crucial for customer service professionals. After all, customers who reach out to support are often confused and frustrated. Being listened to and handled with patience goes a long way in helping customers feel like you are going to alleviate their current frustrations. It is not enough to close out interactions with customers as quickly as possible. Your team must be willing to take the time to listen to and fully understand each customer's problems and needs.
- *Attentiveness*: The ability to truly listen to customers is crucial to providing great service for a number of reasons. Not only is it important to pay attention to individual customers' experiences, but it is also important to be mindful and attentive to the feedback that you receive at large. For instance, customers may not be saying it outright, but perhaps there is a pervasive feeling that your software's dashboard is not laid out correctly. Customers are not likely to say, "Please improve your _X," but they may say things like, "I can never find the search feature" or "Where is (specific function), again?" You must be attentive to pick up on what customers are telling you without directly saying it.
- *Emotional intelligence*: A great customer support representative knows how to relate to anybody, but they are especially good with frustrated people. Instead of taking things personally, they intuitively understand where the other person is coming from, and they know to both prioritize and swiftly communicate that empathy. Think about it: How often have you felt better about a potential grievance simply because you felt immediately heard by the other person involved? When a support representative can demonstrate sincere empathy for a frustrated customer, even just by reiterating the problem at hand, it can help to both placate (the customer feels heard) and actively please (the customer feel, validated in their frustration).
- *Clear communication skills*: Your customer support team is on the front lines of problem solving for the product itself, and serves as a kind of two-pronged bullhorn. On one side, they will be the voice of your company to your customers. That means they must have a practiced grasp on how to reduce complex concepts into highly digestible, easily understood terms. On the other hand, they will represent the needs and thoughts of customers to your company. For example, it does not behoove the customer to receive

a long-winded explanation on the ins and outs of solving a particular bug. The ability to communicate clearly when working with customers is a key skill because miscommunications can result in disappointment and frustration through misinformation. The best customer service professionals know how to keep their communications with customers simple and leave nothing to doubt.

- *Writing skills*: Good writing means getting as close to reality as words will allow. Without an ounce of exaggeration, being a good writer is the most overlooked, yet most necessary, skill to look for when it comes to hiring for customer support. Unlike face-to-face (or even voice-to-voice) interactions, writing requires a unique ability to convey nuance. How a sentence is phrased can make the difference between sounding kind of like a jerk ("You have to log out first") and sounding like you care ("Logging out should help solve that problem quickly!"). Good writers also tend to use complete sentences and proper grammar – qualities that subtly gesture toward the security and trustworthiness of your company. Even if your company offers support primarily over the phone, writing skills are still important – especially when a survey document is prepared. Not only will they enable your team to craft coherent internal documentation, but they also signify a person who thinks and communicates clearly.
- *Creativity and resourcefulness*: Solving the problem is good, but finding clever and fun ways to go the extra mile – and wanting to do so in the first place – is even better. It takes a flamboyant confidence of style or manner to infuse a typical customer service exchange with memorable warmth and personality, and finding a customer service rep who possesses that natural zeal will take your customer service out of "good enough" territory and straight into "tell all your friends about it" land. In other words, you want to have somebody who you do not have to give a lot of rules and regulations to follow. You want to have somebody who is talking to a customer and understands that their boss is really yelling at them today. This person is having a really bad day, and that recognition needs some extra attention to brighten things up. Unfortunately, this is not really something you can teach. The individual must go the extra mile naturally. Very few do!
- *Persuasion skills*: Oftentimes, support teams get messages from people who are not looking for support – they are considering purchasing your company's product. In these situations, it helps to have a team of people with some mastery of persuasion so they can convince interested prospects that your product is right for them (if it truly is). It is not about making a sales pitch in each email, but it is about not letting potential customers slip away because you could not create a compelling message that your company's product is worth purchasing!
- *Ability to use positive language*: Effective customer service means having the ability to make minor changes in your conversational patterns. This can truly go a long way in creating happy customers. Language is a crucial part of persuasion, and people (especially customers) create perceptions about you and your company based on the language that you use. For example,

let us say a customer contacts your team with an interest in a particular product, but that product happens to be back-ordered until next month. How would you respond? Hopefully, in a positive manner, as this will greatly affect how the customer hears the response. There are two options: (a) Responding without positive language: "I can't get you that product until next month; it is back-ordered and unavailable at this time," and (b) Responding with positive language: "That product will be available next month. I can place the order for you right now and make sure that it is sent to you as soon as it reaches our warehouse." The first example is not negative per se, but the tone it conveys feels abrupt and impersonal and could be taken the wrong way by customers – especially in email support when the perception of written language can skew negative. Conversely, the second example is stating the same thing (the item is unavailable), but it focuses on when and how the issue will be resolved instead of focusing on the negative.
- *Product knowledge*: The best customer service professionals have a deep knowledge of how their companies' products work. After all, without knowing your product from front to back, they will not know how to help when customers run into problems. To prevent this ignorance, train the appropriate employees on customer support during their first or second week on the job; it is a critical component of your employee onboarding process. This training will affirm the concepts of CS, the fundamental product/service knowledge, as well as "how" to navigate complex issues.

MITIGATING GAPS IN PRODUCT KNOWLEDGE

To be sure, practically all organizations claim that they have and cultivate CS. However, CS is always changing (evolving) with improvements because each organization sees it as a different entity within the organization. It is not uncommon for different departments to view and interpret CS as different for each other. However, no matter how it is viewed, if CS is seen as "cost-cutting race to the bottom," the results will be devastating for that organization. It takes time for team members to build up their product knowledge. And if you have a very complex product, it may take your team members years to learn every one of its ins and outs. However, the right customer support tool can help you mitigate those gaps in product knowledge. What follows are some thoughts/ideas to facilitate an active and productive CS in any organization. Some of the materials are adapted from https://www.helpscout.com/75-customer-service-facts-quotes-statistics/.

- *Acting skills*: Sometimes your team is going to come across people whom you will never be able to make happy. Situations outside of your control (such as a customer who is having a terrible day) will sometimes creep into your team's usual support routine. Every great customer service professional needs basic acting skills to maintain their usual cheery persona in spite of dealing with people who are just plain grumpy.
- *Time management skills*: On the one hand, it is good to be patient and spend a little extra time with customers to understand their problems and needs.

On the other hand, there is a limit to the amount of time you can dedicate to each customer, so your team needs to be concerned with getting customers what they want in an efficient manner. The best customer service professionals are quick to recognize when they cannot help a customer, so they can quickly get that customer to someone who can help.

- *Ability to read customers*: It is important that your team understands some basic principles of behavioral psychology to read customers' current emotional states. Always bring yourself to the level of customer knowledge. If you cannot do that, chances are you will not be able to communicate. For example: If you talk to them as an academician, they will not take you seriously. On the other hand, if you talk "down" to them, you demonstrate disrespect. So, try to have a balance in your communication ability. Therefore, look for subtle clues about a customer's current mood, patience level, personality, *etc.*, which goes a long way in keeping customer interactions positive.
- *Unflappability*: There are a lot of metaphors for this type of personality – "keeps their cool," "staying cool under pressure," and so on – but it all represents the same thing: The ability of some people must stay calm and even influence others when things get a little hectic. A successful customer service representative knows that they cannot let a heated customer force them to lose their cool. In fact, it is their job to try to be the "rock" for customers who think the world is falling apart because of their current problems.
- *Goal-oriented focus*: Many customer service experts have shown how giving employees unfettered power to "wow" customers does not always generate the returns many businesses expect to see. That is because it leaves employees without goals, and business goals and customer happiness can work together without resulting in poor service. Relying on frameworks like the Net Promoter Score can help businesses come up with guidelines for their employees that allow plenty of freedom to handle customers on a case-to-case basis, but also leave them priority solutions and "go-to" fixes for common problems.
- *Ability to handle surprises*: Sometimes, customers are going to throw your team curveballs. They will make a request that is not covered in your company guidelines or react in a way that no one could have expected. In these situations, it is good to have a team of people who can think on their feet. Even better, look for people who will take the initiative to create guidelines for everyone to use in these situations moving forward.
- *Tenacity*: Call it what you want, but a great work ethic and a willingness to do what needs to be done (and not take shortcuts) is a key skill when providing the kind of service that people talk (positively) about. The most memorable customer service stories out there – many of which had a huge impact on the business – were created by a single employee who refused to just follow the standard process when it came to helping someone out.
- *Closing ability*: Being able to close with a customer as a customer service professional means being able to end the conversation with confirmed customer satisfaction (or as close to it as you can achieve) and with the

customer feeling that everything has been taken care of (or will be). Getting booted before all their problems have been addressed is the last thing that customers want, so be sure your team knows to take the time to confirm with customers that each and every issue they had was entirely resolved.

- *Empathy*: Perhaps empathy – the ability to understand and share the feelings of another – is more of a character trait than a skill. But since empathy can be learned and improved upon, we would be remiss not to include it here. In fact, if your organization tests job applicants for customer service aptitude, you will be hard-pressed to look for a more critical skill than empathy. That is because even when you cannot tell the customer exactly what they want to hear, a dose of care, concern, and understanding will go a long way. A support representative's ability to empathize with a customer and craft a message that steers things toward a better outcome can often make all the difference.
- *A methodical approach*: In customer service, haste makes waste. Hiring deliberate, detail-oriented people will go a long way in meeting the needs of your customers. One, they will be sure to get to the real heart of a problem before firing off a reply. There is nothing worse than attempting a "solution," only to have it miss the mark entirely on solving the actual issue. Two, they will proofread. A thoughtfully written response can lose a lot of its problem-solving luster if it is riddled with typos. Three, and this one may be the most important, it means they will regularly follow up. There is nothing more impressive than getting a note from a customer service representative saying, "Hey! Remember that bug you found that I said we were looking into? Well, we fixed it." That is a loyal, lifetime customer you have just earned. An important side note: The best hires can maintain their methodical grace under regular fire. Since the support team is often tasked with the tough work of cleaning up other people's messes, it is especially important they understand how not to internalize the urgency – and potential ire – of frustrated customers. Instead, they know how to keep a cool head and a steady, guiding hand.
- *Willingness to learn*: While this is probably the most general skill on this list, it is also one of the most important. After all, willingness to learn is the basis for growing skills as a customer service professional. Your team members must be willing to learn your product inside and out, willing to learn how to communicate better (and when they are communicating poorly), willing to learn when it is okay to follow a process – and when it is more appropriate to choose their own adventures. Those who do not seek to improve what they do – whether it is building products, marketing businesses, or helping customers – will get left behind by the people who are willing to invest in their own skills.

What if you are leading a team of support professionals who aren't open to improving their approach to customer service? What if they lack the skills above and do not seem to be interested in developing them?

Often, the root cause of what could be perceived as a lack of skill or unwillingness to learn is the result of a work environment (current or prior) that did not reward going above and beyond to provide excellent service. In this case, try providing your team with some clear guidelines for what you expect and some examples of what great customer service looks like at your company in a way that brings to bear all these skills, and as you do it, make sure that you are celebrating those small wins as you see people starting to use these skills. Once your team starts to see that their efforts are being acknowledged and rewarded, you will have people start to get more engaged, and you will have a clearer picture of whether or not there are actually people on your team who have real skill gaps that you need to work on.

2 Characteristics of Customer Service

OVERVIEW

To set the stage for the anthology, we begin with the notion of the return on investment (ROI), since all organizations focus on it, especially on the quarterly earnings. Customer service (CS), when done well, has just as much impact as a strong sales funnel or excellent marketing efforts. Not only that, but if your customer service team does a fantastic job, it may help boost your other teams' effectiveness. It is easier to make a sale when your customers feel supported, and word-of-mouth marketing after an extraordinary experience is one of the best pitches on the market.

This is emphasized by an alleged statement attributed to Jeff Bezos, who said: "We see our customers as invited guests to a party, and we are the hosts. It's our job every day to make every important aspect of the customer experience a little bit better." The message from Mr. Bezos is loud and clear. Focusing on the customer makes a company more resilient.

Since I am suspicious that you may be a believer of the power and influence of CS in the organization, let us address specific research to convince you. After all, numbers do not lie. Only manipulators of facts and handlers of numbers do.

- A total of 68% of consumers say they are willing to pay more for products and services from a brand known to offer good customer service experiences (https://cdn2.hubspot.net/hubfs/2771217/Content/2018%20Customer%20Service%20Expectations%20Gladly.pdf).
- For 86%, good customer service turns one-time clients into long-term brand champions. (https://khoros.com/blog/must-know-customer-service-statistics).
- An NPS Promoter score has a customer lifetime value that's 600%–1,400% higher than a Detractor (https://www.bain.com/insights/are-you-experienced-infographic/).
- A total of 89% of consumers are more likely to make another purchase after a positive customer service experience (https://c1.sfdcstatic.com/content/dam/web/en_us/www/documents/research/salesforce-state-of-the-connected-customer-4th-ed.pdf).
- A total of 93% of customers are likely to make repeat purchases with companies that offer excellent customer service (https://cdn2.hubspot.net/hubfs/2771217/Content/2018%20Customer%20Service%20Expectations%20Gladly.pdf).
- If the company's customer service is excellent, 78% of consumers will do business with them again after a mistake (https://c1.sfdcstatic.com/content/

DOI: 10.1201/9781003650614-2

dam/web/en_us/www/documents/research/salesforce-state-of-the-connected-customer-4th-ed.pdf).

- Increasing customer retention rates by just 5% can increase profits by between 25% and 95% (https://www.bain.com/insights/are-you-experienced-infographic/).
- Nearly three out of five consumers report that good customer service is vital for them to feel loyalty toward a brand (https://d1eipm3vz40hy0.cloudfront.net/pdf/cxtrends/cx-trends-2020-full-report.pdf).
- A good customer service experience heavily impacts recommendations. Consumers who rate a company's service as "good" are 38% more likely to recommend that company (https://success.qualtrics.com/rs/542-FMF-412/images/ROI%20of%20customer%20experience%202020.pdf).
- Investing in new customers is between 5 and 25 times more expensive than retaining existing ones (https://www.invespcro.com/blog/customer-acquisition-retention/).
- A total of 83% of customers agree that they feel more loyal to brands that respond and resolve their complaints (https://khoros.com/blog/must-know-customer-service-statistics).
- Businesses can grow revenues between 4% and 8% above their market when prioritizing better customer service experiences (https://www.bain.com/insights/are-you-experienced-infographic/).
- A total of 80% of companies use customer satisfaction scores to analyze customer experience and improve it (https://hbr.org/2010/07/stop-trying-to-delight-your-customers).
- A total of 72% of companies believe they can use analytics reports to improve the customer experience (https://www.insightsquared.com/blog/#Switching).
- The customer experience management market worldwide is worth as much as $7.6 billion in 2020. This is a 16.9% year-over-year increase from its value of $6.5 billion in 2019 (https://www.grandviewresearch.com/industry-analysis/customer-experience-management-market). [a. note: The Digital Transformation Initiative of the World Economic Forum expects platform-driven interactions to account for almost 2/3rd of the USD 100 trillion digitalization economy by 2025. As such, the continued integration of Artificial Intelligence (AI) and Machine Learning (ML) into CEM solutions is expected to drive industry growth. Understanding customer behavior is essential for businesses of all sizes. Given that the general public frequently spends quality time on various social networking platforms, such as Twitter now X, Facebook, Pinterest, LinkedIn, and Snapchat, among others, organizations have recognized that social networking platforms can provide a great opportunity to gauge changing customer behavior and the way customers interpret information about products or services. Social customer relationship management (CRM) can particularly assist businesses at this stage in positively engaging customers and increasing brand awareness.]
- As you can tell, excellent customer service is a name maker, no matter what your industry. It helps grow revenue, maintain customer loyalty, and improve your overall business strategy.

- Key customer insights like customer satisfaction score and NPS (Net Promoter Score) help your customer service team improve and grow their impact. Beyond that, though, they also help your other customer-facing teams understand where they could be making improvements. Who would not want to give everyone else a leg up? [a. note: NPS is a score that measures customer loyalty by looking at their likelihood of recommending a given business. To calculate your NPS, subtract the percentage of Detractors from the percentage of promoters. NPS = % promoters – % detractors. For example, if 50% of respondents are promoters, 10% are detractors, and 40% are passives, your NPS would be 50–10 = 40. If your score is 25 or even below, it is considered as a bad NPS score. You need to convert your detractors into promoters.]

THE COST OF BAD CUSTOMER SERVICE

Now that we know how important good customer service can be for your business, it is time to get a handle on the impact it can have when things go badly. Just like good customer service leads to your team's success, poor customer service sets them back. A good experience means that customers are much more inclined to go easy if another experience goes awry. The following customer service statistics help to illustrate this:

- It takes 12 positive customer experiences to make up for one negative experience. (Ruby Newell-Legner's "Understanding Customers") [a. note: It is also worth browsing the statistics of customer experience in the sight: https://businessoverbroadway.com/wp-content/uploads/2013/11/Keepify_CustomerLoyalty_Infographic.pdf.]
- A total of 65% of customers said they have changed to a different brand because of a poor experience (https://khoros.com/blog/must-know-customer-service-statistics).
- After more than one bad experience, around 80% of consumers say they would rather do business with a competitor (https://d1eipm3vz40hy0.cloudfront.net/pdf/cxtrends/cx-trends-2020-full-report.pdf).
- Only one in five consumers will forgive a bad experience at a company whose customer service they rate as "very poor." On the other hand, nearly 80% will forgive a bad experience if they rate the service team as "very good" (https://success.qualtrics.com/rs/542-FMF-412/images/ROI%20of%20customer%20experience%202020.pdf).
- A total of 78% of customers have backed out of a purchase due to a poor customer experience (https://ww2.glance.net/wp-content/uploads/2015/07/Counting-the-customer_-Glance_eBook-4.pdf).

Look at that last one! ***More than three-quarters of customers surveyed have backed out of a purchase because the customer service was not as good as expected***. If you showed that metric to your sales team, they would advocate for your customer service team to get as many resources as they need. Beyond that, customer service sets the stage for the rest of your company to succeed. Your customer service's impact is more widely felt than that of any other team. When they fail, everyone does – and your customers know it.

TYPES OF BAD CUSTOMER SERVICE (AND HOW TO AVOID THEM)

Bad things happen to good intentions. However, by knowing the situations where bad things may happen, we can plan mediating actions to minimize or even eliminate them. We start by recognizing some underlining suppositions and understanding the relevant statistics. They are:

1. *Caring about what customers think*: It is an ill-founded but common misconception that your customers do not know as much as you do. While they certainly are not always right, they are very often at least thinking in the right direction. Similarly, they usually know when something has gone sideways or is not behaving the way it should be.
2. *Your customers' thoughts and opinions matter.* They can help you build a better product, help improve your internal processes, and give you a good heads-up that they are thinking about churning. Here are some customer service statistics that will help you understand just what your average customer feels during or after an interaction:
   - A total of 90% of customers rate an "immediate" response as essential or very important when they have a customer service question. Sixty percent of customers define "immediate" as 10 minutes or less (https://blog.hubspot.com/sales/live-chat-go-to-market-flaw).
   - More than 70% of consumers believe that companies should collaborate on their behalf so they do not have to repeat information to different representatives (https://d1eipm3vz40hy0.cloudfront.net/pdf/cxtrends/cx-trends-2020-full-report.pdft.pdf).
   - A total of 90% of consumers worldwide consider issue resolution as their most crucial customer service concern (https://kpmg.com/xx/en/home/insights/2020/01/customer-first-insights-home.html).
   - A total of 70% of customers report that technology makes it simple to take their business to a competitor if needed (https://www.salesforce.com/blog/category/retail/).
   - A total of 71% of consumers (age 16–24) believe that a quick response from a service team can drastically improve their customer experience (https://www.comm100.com/resources/infographic/millennials-prefer-live-chat-speed-convenience/).
   - Almost 90% of customers report trusting a company whose service they rate as "very good." On the other hand, only 16% of those who give an "inferior" rating trust companies to the same degree (https://success.qualtrics.com/rs/542-FMF-412/images/ROI%20of%20customer%20experience%202020.pdf).
   - A total of 70% of the customer's journey is based on how the customer feels they are being treated (https://www.industryanalysts.com/12417_greatamerica/).

- A customer is four times more likely to switch to a competitor if the problem they are having is service-based (https://www.bain.com/insights/topics/loyalty).
- Nearly 90% of consumers trust a company whose service they have rated as "very good" to take care of their needs (https://success.qualtrics.com/rs/542-FMF-412/images/ROI%20of%20customer%20experience%202020.pdf).

From these statistics, one may surmise that there are at least two metrics that stand out.

1. The first is that most customers indicate that getting a response within 10 minutes of their request is essential or very important – that is a hugely impactful metric. In a world where first response and resolution times are critical, how close are you to that 10-minute mark? With this level of response rate, 71% of younger consumers believe that their experience is drastically improved – and it snowballs from there. When you have created an excellent customer service experience, you gain trust, protect yourself from customers going to your competitors, and prioritize making your customers' journeys good ones.
2. The second exciting statistic is that more than 70% of consumers believe that companies should collaborate on their behalf. So, if a customer reaches out about an issue with an integration that they are using with your product, rather than advising them to reach out to the other product's support team, you respond directly instead.

The conclusion of all these statistics about CS is that the level of personalization and attention to detail can significantly impact your customers and will benefit your profitability in the long run.

NOW LET US IDENTIFY BAD CS

Bad customer service can be defined as when a business fails to meet the customer's expectations in terms of service quality, response time, or overall customer experience. The Poor service negatively impacts the business in several areas, including but not limited to:

- Company and brand reputation.
- Customer churn or decreased customer loyalty.
- Unhappy customers tell friends, families, and colleagues about their experiences.
- Costs associated with loss of revenue, product returns, *etc.*
- Negative online reviews.
- Low employee morale and higher agent turnover.

As bad as all these sound, there is a silver lining in the picture. That is, if the organization has identified that CS is a problem of the organization, they have the opportunity to turn it into a great, successful, productive, and positive experience (opportunity). Obviously, you may not eliminate 100% of the bad customer service, but for sure, you may increase the satisfaction of both customers and employees. You may want to start with prioritizing the biggest offenses, and then follow with a strong focus on the relatively easy fixes, *e.g.*, training, coaching, technology, and data. Table 2.1 shows a microcosm of bad examples based on the list from https://www.playvox.com/blog/bad-customer-service-examples/.

TABLE 2.1
A Microcosm of Bad Examples

	Bad Examples of CS	Corrective Action
1	**Ignore/Trivialize the Concern of the Customer**	**Fix-up**
	It is no surprise that there is a mismatch between customer expectations and what is delivered on the customer service end. From the same Price Waterhouse Coopers (PWC) survey, "only 38% of U.S. consumers say the employees they interact with, understand their needs; 46% of consumers outside the U.S. say the same." While any organization is likely getting feedback from customers ranging from constructive to rage, other dissatisfied customers simply go away and stop doing business. As contact center leaders, are you listening to and acting upon customer feedback? Do you have the data to make impactful improvements and deliver exceptional customer service?	Your team has the power to turn a customer's problem into a positive interaction that builds satisfaction and loyalty. This can sometimes feel like solving a puzzle that is missing pieces. So, all organizations must make the most from customer feedback; however, they receive it: There is no doubt that (a) data exist and (b) what do you do with them? So, how do you start? By analyzing and acting on the available data. Do not wait for more data. The best strategy is: to do the best you can with what you have, and always keep your eyes on changes, so you can improve and adopt to the changes that the continual analysis of the data will show you: If your decision is not based on data, chances are you will fail on your pending changes for improvement. Trust the data, even though sometimes the results may be contrary to the intuition and or experience. This translates into the following: • What does the data tell you? How you answer this question, will determine your success because no matter what you decide it will be a game-changer. • Follow up on negative feedback directly to turn dissatisfied experiences into happy customers. • Take the time to measure and track customer satisfaction and customer sentiment. • Remember that time when a manager followed up or called back? • Monitor interactions between representatives and customers. Make sure you evaluate and prioritize your findings/solutions to identify opportunities for growth improvement, people coaching, and, of course, methods, and tools to improve customer service interactions and prevent negative customer experiences.

(Continued)

TABLE 2.1 (*Continued*)
A Microcosm of Bad Examples

	Bad Examples of CS	Corrective Action
2	**Displaying Poor Attitude and Rude Behavior**	**Fix-up**
	To be sure, CS representatives have one of the toughest jobs out there. They are protecting and building your company's brand – no matter what they have to deal with, from customers. Everyone has good days and bad days. Your representatives will go through times when they just cannot stand the thought of talking to another customer. It happens! It is the role of contact center managers to help representatives remain professional and polite in every transaction – even when the customer is "challenging," "obnoxious," or there is poor communication. The trick is to find the quickest solution for the customer's problem and remember that there is no need to respond with anger and go tit for tat with the customer.	Just like senior managers are protecting your company, contact center leaders and managers need to protect the front-line representatives and their mental health. Building a supportive work climate and modeling positive language and attitude is the first step. Today's contact center technology can take a lot of manual tasks and time-consuming procedures off leaders' and managers' plates. With more time to invest in the customer support team, they can better connect and communicate with agents to keep engagement and motivation high. Deliver appropriate and applicable training not only for new hires but also provide ongoing refresher courses on best practices for handling unhappy customers. Perhaps one of the most critical – one may say, essential – training agenda is to have representatives to be very familiar with the computer and its system that they use in the interaction with the customer. What is appropriate and applicable? The answer will be based on the data analysis of tracking and measuring performance such as: a) Construct scorecards to measure how warm and welcoming agents are b) Do they greet the caller and introduce themselves by name? c) Do they listen to the customer and identify the issue when it is first presented? d) Do they ask the customer if there is anything else they can help with at the end of the interaction? e) Coach representatives on conflict resolution f) Know when a supervisor needs to step in to help mediate.
3	**Empathy**	**Fix-up**
	Customers expect the CS representative to be on their side and apologize on behalf of the company when a product or service fails to meet their expectations. They want to hear the representative say they are sorry and acknowledge the inconvenience caused, and detail how the issue will be resolved.	While you cannot teach empathy, contact center representatives/leaders can take measures to instill a "can do" positive approach to CS. When hiring representatives, include interview questions that gauge empathy and soft skills. Make sure you set up metrics to measure how emotionally responsive and engaging your agents are.

(Continued)

TABLE 2.1 (*Continued*)
A Microcosm of Bad Examples

	Bad Examples of CS	Corrective Action
	If a customer service representative cannot empathize with them and is running through the script with no emotional engagement, the customer might feel as if there is no admission of responsibility or genuine apology. Understanding and caring about customers' needs and frustrations comes with the job in contact centers. Agents must be able to demonstrate they care about your customers' needs and frustrations. Not being able to show empathy through tone, communication, and other means leads to frustrated customers and worst case, lost customers. Using the right tone and language is just one aspect of a great customer service attitude. Your loyal customers also want to feel as though your representative(s) have the solution for every problem, even if they do not. Customers who encounter representatives saying they do not know how to help them cannot inspire confidence and are likely to feel disillusioned with your overall service.	Train representatives to stay positive and approach issues with *emotional intelligence* (especially when there is a generational difference) to minimize the risk of customers ending interactions with an unpleasant experience.
4	**Waiting Time**	**Fix-up**
	This is one of the most common examples of a bad customer service experience. The prospect of waiting in a phone or a physical queue or for a response online for several minutes (or longer) is enough to inspire a sense of dread in even the most upbeat customer. Research consistently has shown that about 30% of customers expect a response in under 5 minutes by phone. And live chat responses follow similar turnaround expectations, if not quicker. A physical queue line is a little more forgiving, with about 15 minutes.	It is vital to get your queue and response times down as much as you can. But easier said than done when you have limited customer support agents available. The solution? Again, technology to the rescue. Two approaches that may be used to fix the problem are: a) Artificial intelligence (AI)-driven solutions and b) Benchmark meaningful KPIs (Key Process Inputs), then set goals to improve the response time and other key metrics. It is imperative for our modern times to develop service skills through automated workflows, performance dashboards, and other tools that encourage "the" customers to access omnichannel support – a win-win.

(Continued)

TABLE 2.1 (*Continued*)
A Microcosm of Bad Examples

	Bad Examples of CS	Corrective Action
	In an increasingly instant culture of wanting everything yesterday, CS centers need to make sure they do not fall prey to long wait times for response and resolution.	
5	**Assessing CS**	**Fix-up**
	Many companies exercise a "token" CS. They offer the opportunity of communication with only email, or they provide a lengthy list of Q&A format of the most frequent concerns – FAQ. In either case, the waiting time for a response may be quite lengthy, which translates into dissatisfaction. Certainly, one size does not fit all when it comes to delivering outstanding service. The goal of contact centers and their representatives is to make it easy for customers to resolve their problems or get information in an effective, timely, and convenient manner. Whereas the jury is still out on the self-service CS, it is safe to say that some form of immediate defense, such as a robust option (*i.e.*, phone, e-mail, messaging, website, chat, and so on), may be appropriate for deflecting common requests. Customers want to reach you when and how they want to reach you in a fast and convenient manner.	Offer robust FAQs and a knowledge library for an immediate and convenient self-service resource. Leverage technology to resolve issues the first time to keep support volumes down. However, do not dismiss the person-to-person interaction. Build an environment where the right people with the right training and knowledge base are ready to help customers in any channel. Make sure the individuals you choose are knowledgeable about the products, services, and your CS system.
6	**Escalating Problem/Concern to Different or Higher Levels**	**Fix-up**
	Being bounced like a ping pong ball from one representative to another is one of the most familiar poor customer service examples. Not only does this imply your representatives do not know how to solve problems, but also, they not have the appropriate authority to resolve issues, concerns and or problems. It also suggests your entire operations need work too.	A critical success factor in providing outstanding CS is providing first contact resolution or FCR. Customers want their problems solved or questions answered on the first try. If frequent FCR is not happening in your contact center, consider auditing the current transfer call rate to uncover root causes. Is more representative training needed? Will segmentation or specialization help support more complex service requests? Are more representatives needed?

(Continued)

TABLE 2.1 (*Continued*)
A Microcosm of Bad Examples

	Bad Examples of CS	Corrective Action
	When a customer calls with a specific issue, be it of a complex technical nature or something more simple, there should be representatives who specialize in the more challenging areas of support, and they have the authority to resolve issues.	Customer service teams should be empowered with proper tools and training to help customers. To develop a toolbox of appropriate tools with the intent to resolve issues on the first try, a good start may be to begin to use checklists, frequency distribution, Pareto Chart, descriptive statistics, and similar simple approaches to data analysis.
7	**Asking Customer to Repeat the Problem/Concern**	**Fix-up**
	When customers are asked to repeat information, it results in frustration and an unsatisfactory support experience. Customers perceive that their time is not being valued or the representative is not really listening. Being asked to give the same personal details, explain the nature of your problem, or anything else multiple times are all avoidable examples of a frustrating customer experience. Often this need to repeat is a result of multiple or disparate systems (including – sometimes different software) that do not talk to each other. Representative A collects customer information only to find out they need to transfer to representative B. Representative B to C, *etc*. To make things worse, sometimes they all work on a different system and have different authorities to resolve problems, which of course infuriates the customer for the time delay and indecision.	Perhaps the number one priority is to give appropriate (realistic, doable) authority to resolve problems. A good metric would be to resolve issues up to a set dollar amount, say $1,000 or some specific amount. However, make sure it is not too low but also, not too high either. Contact centers should work to consolidate technology solutions and update to modern platforms that carry customer information and representative discussions through all phases of customer transactions, regardless of channel. This also allows contact center leaders to monitor for bottlenecks and strive for continuous improvement to boost customer satisfaction.
8	**Escalate Problem/Concern to Other CS Employees or Customer Support is Ineffective**	**Fix-up**
	Many times, a CS representative seems that is ignorant about the product and the system of the organization that he/she tries to mediate with a frustrated customer because of some specified deficiency. The general response is that the problem is passed out to a new representative or escalating to a higher position representative with the same results.	To improve the effectiveness of the CS representative, there are at least five solutions: 1. Regular system knowledge training: It may be on the job training or on an off-site training if the changes are significant. 2. Regular product knowledge training: Make sure your team is up-to-date on all your products, and this should be a company-wide initiative.

(Continued)

TABLE 2.1 (*Continued*)
A Microcosm of Bad Examples

Bad Examples of CS	Corrective Action
Totally unacceptable. You definitely do not want a customer support team that is clueless. Take a moment and imagine such a situation: you call your insurance company to check on specific issue(s) of your policy. You would expect the agent to quickly access your information and understand it. If they cannot even answer basic questions about your policy, that is a big red flag. A customer service team that does not know what they are doing cannot offer proactive help. This leads to a damaged brand reputation and even more complaints from clients.	3. Sharing crucial information: Keep your entire customer service team in the loop about common customer issues and effective solutions. 4. Promoting open communication: Encourage a culture of transparency within your team to better understand and meet customer expectations and needs. 5. Your organizational problem may be your own policies. Maybe it is time to look closely at each policy and see if it is helping or hindering the relationship with customers. We should be open to changing things if it means making our customers' experiences better. An easy approach to evaluate and recommend changes to your policies for CS, you may want to consider the following: • Check each policy closely. If it is making things harder for customers, let us change it or get rid of it. • We should choose policies that put our customers first. Policies should make things smoother, not harder. • Give our team the freedom to make decisions that benefit the customer and the company, even if it means bending some rules. • Keep rethinking – and possibly change – your policies. What worked yesterday might not be the best approach today. Let us stay adaptable and responsive to your customers' needs.

So, in the final analysis, **No** business can afford to underestimate the importance of positive customer experience (CX). Instances of bad customer service will happen and can be detrimental to your business. However, with some initiative from the leaders of any organization, understanding and spotting bad customer service examples from your own contact center is a first step, but the real solution is how quickly and comprehensively you fix the root causes. Certainly, never solve a problem with Band-Aids© as a policy. Obviously, in some situations it will be necessary for such an action, if it is not the official policy of fixing problems. Let us always have in mind that bad customer service can be defined as a (it may also be more than one) failure to meet the customer's expectations in terms of service quality, response time, or overall customer experience. Therefore, to minimize the negative results of poor CS, improve at least the following. The order will depend on the urgency and appropriateness of the situation:

- Lack of empathy.
- Rude customer service.
- Difficult to reach.
- Keeping the customer waiting.
- Not using the right channels.
- Poorly trained or uninformed representatives.
- Lack of resolution.
- Lack of human contact.

3 Statistics Relating Customer Service and Satisfaction

CUSTOMER FEEDBACK: WHY IT IS IMPORTANT

There is something to be said about the power of a personal touch. It has been said by many and quite often that: If a customer likes you – and continues to like you – they will continue to do business with you. If they do not, they will not. Everyone wants to feel known and wanted. They want to believe that the people in their lives are thinking about them, considering them, and understanding their wants and needs. And this, of course, extends to your customers, too. If you go a bit above and beyond to make your customers feel seen, heard, and cared about, it can have a massive impact on their experience and your company's bottom line.

If you can educate yourself on the user's background and history before a call, integrate that into the customer service session. If you are not diving deep and reviewing things like sales history or personal information, at least take the time to ask the customer's name. Read on to find some jaw-dropping customer service statistics around personalization:

- A total of 70% of customers say service agents' awareness of sales interactions is fundamental to keeping their business (https://www.salesforce.com/form/pdf/state-of-the-connected-customer-2nd-edition/?d=7010M000000uQVWQA2).
- A total of 80% of customers say the experience a company provides is just as necessary as its products or services (https://www.salesforce.com/form/pdf/state-of-the-connected-customer-2nd-edition/?d=7010M000000uQVWQA2).
- A total of 63% of consumers expect businesses to know their unique needs and expectations, while 76% of B2B buyers expect the same thing (https://c1.sfdcstatic.com/content/dam/web/en_us/www/documents/research/salesforce-state-of-the-connected-customer-4th-ed.pdf).
- A total of 59% of customers believe that companies need to provide cutting-edge digital experiences to keep their business (https://www.salesforce.com/form/pdf/state-of-the-connected-customer-2nd-edition/?d=7010M000000uQVWQA2).
- A total of 90% of people are more likely to trust a company if they have a firm privacy policy (https://www.salesforce.com/form/pdf/state-of-the-connected-customer-2nd-edition/?d=7010M000000uQVWQA2).

DOI: 10.1201/9781003650614-3

- Your online conversion rate can improve by roughly 8% when you include personalized consumer experiences (https://business.trustpilot.com/reviews/why-a-personalized-consumer-experience-matters-in-the-ecommerce-world).
- A total of 88% of people trust companies that vow not to share their personal information without permission (https://www.salesforce.com/form/pdf/state-of-the-connected-customer-2nd-edition/?d=7010M000000uQVWQA2).
- A total of 92% of customers appreciate companies giving them control over what information is collected about them (https://www.salesforce.com/form/pdf/state-of-the-connected-customer-2nd-edition/?d=7010M000000uQVWQA2).
- A total of 23% of businesses use social media as a tool to collect and analyze data (https://www.gartner.com/en/newsroom).
- A total of 79% of customers are willing to share relevant information about themselves in exchange for contextualized interactions in which they are immediately known and understood (https://www.salesforce.com/form/pdf/state-of-the-connected-customer-2nd-edition/?d=7010M000000uQVWQA2).
- A total of 56% of customers do not mind sharing their personal information in exchange for better service (https://www.salesforce.com/form/pdf/state-of-the-connected-customer-2nd-edition/?d=7010M000000uQVWQA2).
- On average, customer service agents only ask for a customer's name 21% of the time (https://ww2.glance.net/wp-content/uploads/2015/07/Counting-the-customer_-Glance_eBook-4.pdf).

As you can see, it is not just about how you use a customer's personal information, although that is important. Many customers care about how you are aggregating data and what you do with it after you have it. More than half of customers are comfortable sharing personal information in exchange for excellent service, but you still need to use that data carefully. After all, about 90% of customers are more likely to trust you if you have and follow a stellar privacy policy. Of course, personalization helps to build deeper relationships, but only if it is done correctly. If it is not done correctly, you can expect your customers to talk about it.

A SIMPLE MODEL FOR AN EXCEPTIONAL CUSTOMER SERVICE

- Differentiate care, and experience, every time.
- Highest performing network.
- World–class service lines.
- Local market leadership.
- Transformative partnership.
- Inspire your employees through an extraordinary place to work.

The focus is the center of who we are trying to please with their wants and needs. That is: THE CUSTOMER (one person at a time)! So, in essence, the attention for an exceptional customer service is:

- Focus on a safe environment and best outcomes. (Remember that the customer with a specific or a given complaint may be aggressive. However, the unhappy customer opens an opportunity to see where the issues exist in your organization, which the satisfied customer never gives you.)

- Provide an exceptional experience. This is accomplished by having a philosophy of *Gradatim Ferociter* or "Step by Step, Ferociously." Exceptional experiences grow over time. They never appear without effort and perseverance.
- Show empathy for the customer's concern(s), compassion for the anxiety that s/he may display, and certainly show genuine interest in resolving the specific problem. This means that the customer service your organization offers is committed to customer satisfaction and a true commitment to resolving "the" conflict at hand. After all, let us realize that the customer may be angry!
- Provide an affordable and efficient resolution, so that the values of your organization are demonstrated with a given transaction. Those values may be: compassion, innovation, and respect for the customer. Of course, for your organization, there may be others.

THE IMPORTANCE OF WORD-OF-MOUTH

Customers love to talk. They will share their good experiences, bad experiences, and everything in between. Similarly, customers like to listen.

As you will see in the following customer service statistics, many users trust reviews from other customers or social media responses more than they trust the marketing of your actual company.

Providing an excellent customer service experience, especially within the public eye of social media, is a massive boon to your company's marketing efforts. While 50% of customers do not share their experiences on social media, 72% will talk about them in person. That number gets even higher when considering how customers talk socially about poor experiences. Read on for some pretty impressive and shocking customer service statistics around when people do or do not share:

- A total of 36% of consumers will share their customer service experience, whether good or bad. More than one-third report posting on Facebook, followed closely by Instagram (https://cfigroup.com/wp-content/uploads/CFI-contact-center-satisfaction-2020.pdf).
- A total of 87% of consumers read online reviews for local businesses in 2020 (https://www.brightlocal.com/research/local-consumer-review-survey/).
- A total of 72% of customers will share their good experiences with others (https://www.salesforce.com/form/pdf/state-of-the-connected-customer-2nd-edition/?d=7010M000000uQVWQA2).
- A total of 72% of customers will tell six or more people if they have a satisfying experience (https://www.slideshare.net/slideshow/cx-for-executives/52398185#2).
- A total of 94% of American customers will recommend a company whose service they rate as "very good" (https://success.qualtrics.com/rs/542-FMF-412/images/ROI%20of%20customer%20experience%202020.pdf).

- A total of 67% of customers report a terrible customer experience as the reason for switching businesses (https://www.slideshare.net/slideshow/cx-for-executives/52398185#2).
- Only 13% of consumers will recommend a company whose customer service they have rated as "very poor" (https://success.qualtrics.com/rs/542-FMF-412/images/ROI%20of%20customer%20experience%202020.pdf).
- A total of 50% of customers say they do not share their bad or good service experiences through social media (https://cfigroup.com/wp-content/uploads/CFI-contact-center-satisfaction-2020.pdf).
- A total of 33% of consumers would recommend a brand that provides a quick but ineffective response (https://hbr.org/2012/12/turn-customer-care-into-social).
- A total of 78% of customers who complain to a brand on Twitter expect a response within an hour (https://cfigroup.com/wp-content/uploads/CFI-contact-center-satisfaction-2020.pdf).
- A total of 13% of customers tell 15 or more people if they have a negative experience (https://www.slideshare.net/slideshow/cx-for-executives/52398185#2).
- A total of 79% of customers who share their complaints online see their complaints ignored (https://www.slideshare.net/slideshow/2011-customer-experience-impact-report/10947124#4).
- Only 17% of consumers would recommend a brand that provides a slow but effective solution (https://hbr.org/2012/12/turn-customer-care-into-social).
- Only 1 in 26 customers will tell a business about their negative experience; according to customer service facts, the rest simply leave (https://www.slideshare.net/slideshow/cx-for-executives/52398185#8). [a. note. That statistic about customers leaving without a word horrifies us every time! Consider how many negative responses you have had to CSAT (Customer satisfaction score is a commonly used metric that indicates how satisfied customers are with a company's products or services), or NPS surveys (Net Promoter Score, which is a common metric used in customer experience programs. An NPS score measures customer loyalty by looking at their likelihood of recommending a given business) then multiply it by 26 – scary, right? Luckily, some great current technologies make it a bit more straightforward to provide an excellent customer support experience and mine for nuggets of valuable insights as you do it.]

SOME TECHNIQUES FOR COLLECTING VOICE OF THE CUSTOMER DATA

Data is as good as the selection process. Therefore, choosing the right support channels for selection is of paramount importance. Both phone and email are the tried-and-true channels of customer service. Back in the 1960s, when call banks first came about, customers started getting comfortable with having a direct line to support. Then came email, providing either a single address or a group email to reach customer service teams.

TABLE 3.1
Selected Channels for Receiving Feedback

Selected Channels for Receiving Feedback

	Channel	Used by Service Teams	Used by Customers
1	E mail	95%	98%
2	Phone	95%	93%
3	In Person	89%	94%
4	Social media	72%	71%
5	Online forms	68%	83%
6	Knowledge base	66%	82%
7	Customer Portals	64%	84%
8	Text/SMS	63%	78%
9	Messager apps	55%	72%
10	Online chat/live support	52%	81%
11	Mobile app	51%	82%
12	Online communities/discussion forums	48%	68%
13	Mobile chat	43%	N/A
14	Video support	33%	N/A
15	Video-activated personal assistants	20%	54%

Data represents service decision maker responses.
State of the Connected Customer Salesforce Research June 2018.
Each channel has pros and cons. Here are some customer service statistics to help illuminate the positives and negatives for each and to help with your decision-making.

Now there's live chat, co-browsing, virtual video calls, social media mentions, and tons of newly arriving methods for customers to reach out. Which one is best for your company? Table 3.1 shows some different channels and how often each is used by support teams and customers.

OMNICHANNEL SUPPORT

Omnichannel support is a continuous connection for customers between all of your channels. It effectively means that what someone does with a team on any channel can travel seamlessly to other channels with full context. For instance, your customer can start with a live chat and then transition seamlessly to a phone call as needed.

Omnichannel support provides an excellent experience for customers because it means that they never have to re-explain something they already told another member of your team. When done correctly, it can seem like magic:

- A total of 75% of customers desire a consistent experience, regardless of how they engage a company (through social media, in person, by phone, *etc.*). (https://www.salesforce.com/blog/category/retail/).

- A total of 64% of customers want to shop with companies that can meet their needs in real-time (https://www.salesforce.com/blog/category/customer-experience/).
- A total of 64% of customer service teams use customer portals, compared to 84% of customers who use customer portals (https://www.salesforce.com/resources/research-reports/state-of-service/).

One can see that timeliness, the ability to use multiple channels, and consistency are the main benefits of omnichannel.

4 Service Strategies

OVERVIEW

In today's global world, we operate in a competitive market, no matter what the industry or service organization we belong to. However, no matter where we are, or what we produce or service, the one thing that still routinely separates a successful organization from a failing one is indeed how they handle their customer service. As great customer experiences set a company apart in a crowded marketplace because they create passionate and vocal customers. This competitive advantage is hard for other businesses to break down because they must battle that existing business-to-customer relationship first, and they need to craft a similarly good customer experience.

So, to get ahead of the curve, any organization must establish the best startup customer service strategy or plan and practice it as early as possible. However, crafting a great customer experience requires management leadership to purposefully invest resources in customer service. To deliver the best customer service, especially as a startup, some suggestions are:

1. *Prioritize appropriately*: The best thing you can do to improve your customer service as a startup is to prioritize your tickets. Start by defining these three things:
   - The scope of the pain the customer is experiencing.
   - Whether their issue is widespread or affecting only them.
   - How long they have been waiting.

   Each of these should play into how you prioritize tickets and how quickly they need a response. For high-pain or high-frequency tickets, it is important that they get a response quickly. Then, traditionally, teams answer lower-pain issues in the order that they are received. For issues that are not high-pain or widespread, consider this nontraditional approach: Rather than answer the oldest tickets first, start with the newest. For example, imagine this scenario: 50 emails are waiting in the support inbox this morning. The oldest is from about 12 hours ago, and the newest came in 5 minutes ago. Assume that it will take your team an hour or two to work through this inbox, regardless of which order they take it on. If the person who has already waited 12 hours needs to wait for another hour or two, it does not significantly impact their waiting time. As long as your first response to them is high-quality and helpful, they will likely be satisfied. But suppose your team replies to that customer who emailed 5 minutes ago immediately. In that case, you have the opportunity to create a "wow" experience by resolving an issue within just a couple of minutes.
2. *Recognize and reward customer centricity*: As your company grows, teams will often silo themselves and become less communicative. Be sure to encourage team-wide customer centricity instead of developing deeply

DOI: 10.1201/9781003650614-4

specialized teams, especially early on. This means, on time to pay attention to and reward customer centricity by implementing the following strategies:

- Talk about it in your all-hands meetings, offer rewards for excellence, and build out your Voice of the Customer strategy.
- Give customer service a seat at the table when discussing company-wide strategies and metrics.
- Empower your customer-facing teams to pull in whoever they need to accomplish their goals. Ensure that they work closely with engineering, product, and marketing teams to keep everyone customer-focused.

That said, talking about your customers is only half the battle – you need to encourage customer centricity within the context of customer conversations as well. That means, for instance, engineers prioritizing bug fixes over a more attractive, larger project, or toning down technical language while on a troubleshooting call. For marketing, it might mean making it easy to unsubscribe from an email list or responding to all of the comments customers leave on the blog. Customer centricity goes beyond just your customer-facing teams.

3. *Prioritize using tools that help*: At the beginning of a startup, it can feel like pulling teeth to get a budget for things that are not desperate and dire. However, getting the right tools in place from the start can be game-changing when it comes to your customer service. Identify the tools that will make a big impact, and spend money on those. Do not go overboard, but do not skimp on tools that will help people do their jobs better. For some companies, that may simply be setting up a support-specific email address and designating office hours for team members to work on tickets. When you are small, it is best to keep your process simple, after all. However, as you grow, you may want to find something slightly more scalable. That is where tooling can help.
4. *Do not hack solutions*: Rather than loosely cobbling solutions together, invest in developing the right solutions the first time. Unfortunately, often, when companies are small and strapped for time and cash, they will push roundabout fixes for simple issues to keep their engineers focused. Take refund processing, for instance: Rather than build a refund processing functionality into a support back end, many companies rely on multiple external tools to offer refunds. Without a built-in, back-end refund process, a customer support representative may need to:
   - Go into your billing platform.
   - Get a transaction ID.
   - Go to the company's payment processor.
   - Find the payment.
   - Ensure that it is within the refund period.
   - Manually process the refund.
   - Then respond back to the customer with a refund ID.

   Think about how much time that takes up for just a few refunds. Rather than doing something that saves time in the short run and builds up technical debt, help your team find the right solution the first time.

Strike a balance between getting things done quickly and being able to scale when needed. Calculate the amount of time that a workaround will take (short-term fix) versus the amount of time implementing a more scalable solution will take (long-term fix). This goes for both internal and external features.

5. *Set trackable goals*: You do not know what you need to improve if you do not track it. At the core of every excellent startup, customer service team are meaningful, trackable metrics.

   Instead of just assuming that your customers are happy because they continue to use your product, try to find quantifiable ways to measure your improved customer service delivery.

   Some metrics that serve as great baselines for improvement are:

   - *NPS*: NPS stands for Net Promoter Score, and it is intended to measure customer loyalty. NPS was created by Fred Reichheld and was first discussed in his 2003 HBR article "The One Number You Need to Grow" on the same topic. Reichheld, F. (2006). *The Ultimate Question: Driving good profits and true growth.* Harvard Business School Press. Boston, Massachusetts."
   - *CSAT*: The concept of a customer satisfaction score (CSAT) first gained recognition with the dawning of mass production. Movements such as Henry Ford's mass production of vehicles led to dramatic decreases in production cost and made the importance of individual customer experience much less critical. These days, we have swung back the other direction. Even though mass production is still the norm, customers note the bounty of options at their fingertips and differentiate based on how satisfied they are with the product or service they are receiving.

     CSAT, as we know it now, probably came about in 1994 when the University of Michigan established the American Customer Satisfaction Index (ACSI). It was the first measurement implemented to identify the value of goods and services from the perspective of the customer. Modern CSAT still does just that: It takes an ongoing measurement of the overall satisfaction your customers have with a service, product, or relationship with your company.

     At this point, one may wonder. What is CSAT? Many organizations define it in several ways with different metrics. However, the most common and useful is the *Response time*. That is: How long is too long for a customer to wait for a reply to their question? Does "faster" equal a meaningfully better experience? Typical metrics are:

     - First contact resolution rate.
     - Handle time.
     - Ticket volume.

   Beyond just serving as great individual metrics, when viewed collectively, you can understand correlation and causation. (Caution: They are not the same. The first shows relationship, the second shows causation.) For instance, as ticket volume goes up, does response time go up as well? When handling time drops, does CSAT improve? Understanding how changes can impact your customers' experiences gives you the data-empowered position to make meaningful changes.

6. *Shorten the wait for escalated help*: Help your customers get their problems solved as quickly and painlessly as possible by connecting them with the people who can help them right away.

   Create processes that allow your team to expedite escalations. Rather than a team member spinning their wheels trying to solve an issue that they may not be the best equipped to handle, identify the person who can solve the issue right away. It sounds like common sense, but unfortunately, it happens, and as a result, the customer becomes dissatisfied. When you guide the customer to the person who has both responsibility and accountability for the resolution, it is the best for your customers and best for your team. Everyone is working on conversations that they are equipped to handle, and your customers will move through your customer service processes more smoothly. Typical interactions help in the outcome.

   Provide conversation summaries as part of the escalation process.

   - Once you have figured out the right person to respond to a customer query, consider making conversation summaries part of your handoff routine.

   Including a summary helps you:

   - Avoid having to ask customers to repeat themselves.
   - Research escalated issues more efficiently.
   - Understand the full context of a customer's request without having to sift through long email threads.

   A good AI software will be of great help to create a bulleted summary of any conversation with just one click of a button, even further simplifying the process.

7. *Write excellent documentation*: No matter how many exceptional employees you have working to answer customer inquiries, your documentation is always your first line of defense for customer service, and it can be especially helpful for startups. Beyond information for your customers, documentation also helps your internal teams get onboarded faster and build consistent knowledge of your product more easily. When there is documented information that can be referenced quickly, it is easy for both your customers and your new employees to find answers to questions without needing to ask someone else. It is worth remembering that Statistics suggest that 40% of customers prefer using self-service over reaching out to other customer service channels (see next section for specifics). One of the most effective self-service options is a detailed knowledge base. [Remember the axiom: If it is not documented, it does not exist.]

   Not only does this provide a better customer experience because it gives your customers what they are looking for, but it also helps to deflect support conversations. In startup customer service, the more low-level tickets you can deflect, the more your team will be able to focus on the conversations that need a human touch.

Look at the common tags in your inbox and categorize the topics that are not already included in your documentation. Consider conversations where your team has not linked documentation to your customers and identify whether a new doc or resource might help. Whatever you do, do not limit yourself to only documentation. Instead, consider how multimedia may help: Videos, pictures, and infographics can be a great way to reach folks who do not necessarily glean as much from reading.

8. *Surprise and delight your customers*: Remember that your customers are human first. Often, the best way to help them achieve their goals is with a human connection. No matter how effective a knowledge base or a quick answer is, sometimes an interaction calls for extra surprise and delight. Look at air travel, for example. There are tons of airlines you can choose from when it comes to flying. Some are cheaper than others; some offer better amenities. But Southwest has made a name for itself in the industry. The tickets are not always the cheapest, and the seating arrangement can be anxiety-inducing. However, they care about their customers.

   Southwest accommodates flight changes and cancellations, often without added fees. They are known for singing happy birthday to children on flights, bringing extra amenities to new mothers, and just generally going the extra mile. This type of goodwill cannot be developed through technology; it comes from the humans who invest in customers every day.

   Your team can create similar experiences whenever they are interacting with customers. If they are already logged into a customer's account, they can look for and suggest ways to maximize the value the customer gets from using your product. You can also send swag or thank-you notes to make an impression. It requires some additional time, but the benefit outweighs the cost.

   Help your team find ways to leverage human connection to create moments of surprise and delight. Whatever you implement, the goal should be to establish goodwill with customers by being helpful and responsive and going above and beyond if possible.

9. *Ask your customers for their help – and use it*: When you ask for feedback, make sure that you implement it. Nothing feels more futile than taking the time to provide information and insight and then getting the sense that it is being ignored. Take the opportunity to ask for additional thoughts or actions after you have helped in a customer service interaction. You will not get helpful feedback every time, but it will be worth it when you do. If you have identified something more helpful than customer feedback (a product review, for instance), take the chance to ask that instead. The approach can be simple. You help the customer with their issue, and once you have that resolved, you can ask something from them:

   - With happy customers, you might ask them to write a review of your product.
   - For existing customers, you might offer to help them implement more of your tools.
   - For prospective customers, you might make them an offer to become a paying customer.

- You have your customer's attention at that moment, so use that opportunity to provide an excellent customer service experience and make progress on other goals.
- Leverage what you have. The key to creating excellent startup customer service is to take a "cost plus something" mindset: Try to make a benefit beyond just your input. If time is your exclusive input, you will never have enough. If you can apply that time to create more time in the long term, you will build something that can continue to grow. Leverage what you have today: Focus on the actions that give you the best return on your limited resources, and maximize the value and outcome of every customer interaction.

MANAGING SOCIAL MEDIA CUSTOMER SERVICE: STRATEGIES AND TIPS

Phone support: Phone support is one of the oldest forms of assistance. Since the invention of the rotary phone, companies and customers alike have depended on this valuable, reliable technology. Let us look at some customer service statistics around teams using phones:

- A study has shown that more than 76% of all consumers prefer the traditional medium of phone calls to reach customer support representatives (https://cfigroup.com/wp-content/uploads/CFI-contact-center-satisfaction-index-2019.pdf).
- More than 50% of customers across all age groups typically use the phone to reach out to a service team, making it the most-used channel for customer service (https://d1eipm3vz40hy0.cloudfront.net/pdf/cxtrends/cx-trends-2020-full-report.pdf).
- A total of 63% of customer service teams use text messaging, compared to 78% of customers who use text messaging for communicating with a company (https://www.salesforce.com/resources/research-reports/state-of-service/).
- A total of 51% of customer service teams use mobile apps, compared to 82% of customers who use mobile apps for communicating with a company (https://www.salesforce.com/resources/research-reports/state-of-service/).

Although support teams heavily use traditional phone support, it seems that there are still some tools they could use more. For instance, the number of customers who prefer text messaging and mobile apps compared with the number of customer service teams that use them leaves something to be desired. Phone support is tried and true, but the opportunities available through mobile phones seem to still be a new frontier.

Self-service support: Documentation is beneficial for assisting customers without the need for your customer service team's intervention. For example, a customer can search in your knowledge base and find answers within minutes, which, in the long run, is a much better experience than having to wait for a response in an email or live chat queue. Unfortunately, though, when documentation is not done well, it can do more harm than good. Here are some customer service statistics exploring both the good and bad sides of your help docs:

- A total of 28% of consumers say the most frustrating issue is information that is simple but hard to find (https://www.drift.com/learn/chatbot/why-are-chatbots-important/).
- A total of 69% of consumers first try to resolve their issue independently, but less than one-third of companies offer self-service options such as a knowledge base (https://d1eipm3vz40hy0.cloudfront.net/pdf/cxtrends/cx-trends-2020-full-report.pdf).
- Customers prefer knowledge bases over all other self-service channels (https://www.forrester.com/report/2018-Customer-Service-Trends-How-Operations-Become-Faster-Cheaper-And-Yet-More-Human/RES142291).
- A total of 66% of customer service teams use knowledge bases, compared to 82% of customers who use knowledge bases (https://www.salesforce.com/resources/research-reports/state-of-service/).

Based on these statistics, one can see that most customers would rather use a knowledge base/documentation to resolve their issues, but they are not always able to find them. Few companies have made efforts to create documentation, knowledge bases, or other options with which customers can serve themselves.

Social media support: Despite social media's role as a place to connect with friends and coworkers, many businesses and consumers are using it to connect as well. Functionality like Facebook and Instagram messaging have made it easy for users to connect directly with brands in a more meaningful and authentic way than they used to. Because of that, many companies are striving to create more presence in those environments. These two customer service statistics around social media make it easy to understand why:

- A total of 79% of American internet users are on Facebook, Twitter (now X), making it the most popular social media platform (https://www.lyfemarketing.com/blog/social-media-marketing-statistics/).
- A total of 95% of adults between the ages of 18–34 are likely to follow a brand through social media channels (https://www.marketingsherpa.com/article/chart/demographics-why-customer-follow-brands-social-media).

Think about all of the people your brand could connect with if you were to make the switch to providing customer service via social media channels? [If you are interested in measuring this activity, there is a metric that may help you doing so. It is a tool that is used in marketing (especially in advertising) and is called cost per view (CPV). If you are interested in how much social media is costing you, it is an applicable tool. It is calculated by dividing the total cost of your message production and delivery by the total number of participants.]

Chat support: Chat is primarily considered to be the fastest way for consumers to get support. It feels immediate, can often be fielded by automation, and does not always require a lot of bandwidth from staff. Similarly, many younger consumers prefer it over more traditional technology like phones. If you have not already implemented chat within your team, here are two customer service statistics that may help you understand if it would be a good move:

- Millennials prefer live chat for customer service over every other communication channel (https://www.comm100.com/resources/infographic/millennials-prefer-live-chat-speed-convenience/).
- A total of 52% of customer service teams use online chat or live support, compared to 81% of customers who use online chat or live support for communicating with a company (https://www.salesforce.com/blog/customer-service-stats/).

There are many more customers who prefer to use live chat than companies that offer it! If you have the resources to implement it, chat may be an excellent opportunity to meet your customers where they are. Note that if your customers are searching for live chat and you do not offer it, it may frustrate them even more.

DEALING WITH ANGRY CUSTOMERS

Earl Nightingale, allegedly, has said that "Our attitude towards others determines their attitude towards us." By extension, we can say that our attitudes do indeed contribute to anger. Anger is a widespread and extremely difficult emotion to control. For example, one-third of people who responded to a survey about anger noted that they or someone they knew had a severe problem controlling their anger (https://www.mindyouranger.com/anger/anger-statistics/). It may come as no surprise, then, that many of your customers, who are already in a stressful situation and need help, may become angry during an interaction.

Anger can be scary, but when you understand where it is coming from and why it's happening, it can be easier to handle. For example, many customer service team members could benefit from reminding themselves that the customer is angry about the situation that led them to the interaction, rather than the interaction itself. For a few more reminders about anger within the context of customer service and how certain behaviors can lead to more profound customer anger, check out these customer service statistics:

- A total of 21% of people waiting at a hospital pharmacy decided to fill their prescription elsewhere because they had to wait too long (https://journals.sagepub.com/doi/abs/10.1177/001857870103600307).
- Nearly 70% of customers are irritated when their call is transferred from one department to another (https://d1eipm3vz40hy0.cloudfront.net/pdf/cxtrends/cx-trends-2020-full-report.pdf).
- Feeling unappreciated is the #1 reason customers switch products and services (https://www.vonage.com/contact-centers/).
- A total of 33% of customers are most frustrated by having to wait on hold, and 33% are most frustrated by having to repeat themselves to multiple support reps (https://blog.hubspot.com/sales/live-chat-go-to-market-flaw).
- A total of 70% of unhappy customers whose problems are resolved are willing to shop with a business again (https://ww2.glance.net/wp-content/uploads/2015/07/Counting-the-customer_-Glance_eBook-4.pdf).

- Nearly 60% of customers feel that long holds and wait times are the most frustrating parts of a service experience (https://d1eipm3vz40hy0.cloudfront.net/pdf/cxtrends/cx-trends-2020-full-report.pdf).
- A total of 62% of customers would prefer to "hand out parking tickets" than wait in an automated phone tree for service, or must repeat themselves multiple times to different team members (https://cdn2.hubspot.net/hubfs/2771217/Content/2018%20Customer%20Service%20Expectations%20Gladly.pdf).
- A total of 35% of customers have become angry when talking to customer service (https://yourbusiness.azcentral.com/impact-negative-word-mouth-9567.html).
- A total of 27% of Americans report that ineffective service is their number one customer service frustration (https://www.statista.com/statistics/815599/causes-of-customer-service-frustration-us/).
- A total of 12% of Americans rate their number one service frustration as "lack of speed" (https://www.statista.com/statistics/815599/causes-of-customer-service-frustration-us/).
- A total of 72% of customers say that explaining their problems to multiple people is poor customer service (https://d16cvnquvjw7pr.cloudfront.net/resources/whitepapers/Zendesk_WP_Customer_Service_and_Business_Results.pdf. Key findings here are: (a) Customer service both good and bad impact revenue, (b) Customer service stories are spread widely – especially bad ones, and (c) Social media drives increased sharing of customer service experiences. It is interesting that the participants of this research project specifically ranked customer service as the #1 factor impacting vendor trust. Some of the actual statistics are:
  - A total of 62% of B2B and 42% of B2C customers purchased more after a good customer service experience (B2B = Business to Business).
  - A total of 66% of B2B and 52% of B2C customers stopped buying after a bad customer service interaction.
  - A total of 88% have been influenced by an online service review when making a buying decision:
    - Fast customer service matters.
    - A total of 69% attributed their good customer service experience to the quick resolution of their problem.
      - A total of 72% blamed their bad customer service interaction on having to explain their problem to multiple people
    - Customer service experience(s) has a long-lasting impact.
    - A total of 24% continue to seek out vendors two or more years after a good experience.
      - A total of 39% continue to avoid vendors two or more years after a bad experience.
      - Women (45%), B2B (51%), Gen X (54%), and high-income households (79%) are most likely to avoid vendors two or more years after a bad customer service experience.
- A total of 79% of consumers who used online feedback to complain about a poor customer experience were ignored. Right Now's Annual Research

Shows 86 Percent of U.S. Adults Will Pay More for A Better Customer Experience (https://www.businesswire.com/news/home/20120111005284/en/RightNow%E2%80%99s-Annual-Research-Shows-86-Percent-U.S.).
- A total of 84% of customers report that their expectations had not been exceeded in their last interaction with customer service (https://hbr.org/2010/07/stop-trying-to-delight-your-customers).
- A total of 78% of customers have given up on a transaction because of a negative customer experience (https://www.slideshare.net/slideshow/2011-customer-experience-impact-report/10947124#6).
- A total of 67% of customers end a call in frustration when they cannot reach a customer service representative (https://ww2.glance.net/wp-content/uploads/2015/07/Counting-the-customer_-Glance_eBook-4.pdf).

Most customer frustration seems to stem from waiting or having to repeat themselves. What are some ways you might be able to speed up support interactions or provide additional context for team members to work against these two common issues? Putting some processes in place may help you work on these customer service statistics in the future. Preemptively addressing customer concerns and frustrations before they occur is a mark of a truly high-performing customer service organization.

Give customers a great experience, and they will buy more, be more loyal, and share their experience with friends. Great. That is what every company strives for. So why are so many consumers disappointed? Call it an experience disconnect: Companies tout the latest technology or snappy design, but they have not focused on – or invested in – the aspects of customer experience that are the most meaningful. *What truly makes for a good experience? Willfulness to resolve issue/concern/problem. That is: Speed. Convenience. Consistency. Friendliness. And one big connector: human touch* – that is, creating real connections by making technology feel more human and giving employees what they need to create better customer experiences. People are increasingly loyal to the retailers, products, brands, and devices that consistently provide exceptional value with minimum friction or stress. The challenge: Use new technology with purpose to make the experience feel more human – without creating frustrations for customers and while empowering employees.

BUILDING A WORLD-CLASS SUPPORT TEAM

In an era when companies see online support as a way to shield themselves from "costly" interactions with their customers, it is time to consider an entirely different approach: Building human-centric customer service through great people and clever technology. So, get to know your customers. Humanize them. Humanize yourself. It is worth it. No matter how much we talk about anger, frustration, or disappointment, the fact remains that customer service teams do the brunt of the work to ensure your customers are happy and successful. Unfortunately, customer service is often considered last when thinking about company policies, sharing information, or empowering individuals and/or teams. However, if the statistics have shown anything, it is that customer service is essential and valuable to your customers. Studies have repeatedly shown the best practices that top-of-their-class customer service people put into practice and how it makes their customers feel:

- A total of 75% of consumers will still choose to interact with a real person even as the technology for automated solutions improves (https://www.pwc.com/us/en/advisory-services/publications/consumer-intelligence-series/pwc-consumer-intelligence-series-customer-experience.pdf).
- One-third of all consumers consider the most critical aspect of good customer service experiences to be having their problem solved in one single interaction, regardless of the amount of time spent (https://www.statista.com/statistics/810614/important-aspects-of-a-good-customer-service-experience/).
- A total of 88% of high-performing service decision-makers are investing significantly in agent training compared to only 57% of underperformers (https://www.salesforce.com/content/dam/web/en_gb/www/pdf/salesforce-research-third-edition-state-of-service.pdf).
- A total of 83% of high-performing service agents say they get the training they need to do their job well, compared to only 52% of underperformers (https://www.salesforce.com/content/dam/web/en_gb/www/pdf/salesforce-research-third-edition-state-of-service.pdf).
- A total of 69% of high-performing service agents actively look for situations to use artificial intelligence (AI) compared to only 39% of underperformers (https://www.salesforce.com/content/dam/web/en_gb/www/pdf/salesforce-research-third-edition-state-of-service.pdf).
- A whopping 95% of consumers say that customer service is essential for brand loyalty, and 60% of consumers report having deserted a brand and switching to a rival company because of poor customer service (https://info.microsoft.com/rs/157-GQE-382/images/2018StateofGlobalCustomerServiceReport.pdf).

It seems like the main differentiator for performance is appropriate training and company trust. Of course, it is also imperative for the consumer to notice the difference. When you take care of your employees and ensure that they have the tools they need, they can better care for your customers.

Of course, when adversity or exceptional events happen, they do affect CS. An example of this unusual adversity was the impact of COVID-19. The main trends are that customers are more challenging to please, more willing to switch to a company that aligns with their needs, and significantly less patient. We cannot blame them! However, given these shifts, your customer service team may need to change their strategies to directly address the concerns that customers in this new world have (all the following statistics are from https://hbr.org/2020/04/supporting-customer-service-through-the-coronavirus-crisis):

- The typical level of difficult calls ranges from 10% to 20%, but the pandemic made customer experience statistics worse.
- The average company saw the percentage of calls scored as "difficult" by reps increase by more than 100%.

- There were more unexpected travel cancellations, appeals for bill payment extensions, insurance coverage disputes, and other anxiety-induced calls than at any other historical time.
- Customer service stats from 2020 show that reps feel like their job is much harder and less enjoyable.
- Some of the companies in the study saw hold times grow by 34% and escalations increase by more than 68%, further highlighting the impact customer service has on businesses.
- One of the Tethr AI study companies provided customer service statistics that showed difficult interactions had a 6% chance of resulting in an upsell or cross-sell. The chances are usually 80% when the customer service experience is excellent.
- Only about 4% of customers who leave or threaten to leave are enticed to stay by accepting a promo, down from the average of 20% in the past.

Not only has COVID-19 impacted working environments and employees' emotions, but it has also affected customers in the same way. It is much harder to connect with and solve problems for a customer who is burnt out and frazzled than it would be for a customer who feels secure and content. Unfortunately, COVID-19 created a stressful position, making it infinitely harder to connect on a human level and meet customers' needs through support.

Hopefully, all of those customer service statistics have given us much to consider and process. There are many different perspectives to consider and options to evaluate, but no matter what you have taken away from these statistics, it is sure you will agree with this: *Customer service is not only essential, but it is valuable to your bottom line and serves as a critical differentiator of your company against others. If you treat every interaction as a way to get to know and love your customers, your team will be providing excellent customer service in no time.*

Finally, management must demonstrate that CS matters. One way to do it is through their leadership in at least some of the following principles:

- Take ownership of both successes and failures. If appropriate, give credit to your employees.
- Always focus on innovation and simplicity.
- Always demonstrate obsession toward "the customer," by being curious about their needs and wants.
- Recognize that customers may be right most of the time.
- Set or adopt standards that provide opportunities for CS, productivity, efficiency, and the welfare (growth) of the organization.
- Hire and cultivate the "best" employees and or management team.
- Above all, **think big**. Do not underestimate the *impossible*. It can be tackled with success!

5 Economic Utility and Its Relationship to Customer Service

OVERVIEW

The idea of understanding *Economic Utility* in any organization is fundamentally to create customer value, which, of course, should be the "ultimate goal" of any business. While not every campaign requires the complete implementation of all six utility types to improve conversion and customer satisfaction, general knowledge paves the way for implementation to deliver value at scale.

What does utility mean in economics? Utility theory in economics pertains to the value or worth of a certain good, service, or item. It suggests that goods, services, and items can be ranked according to their usefulness. The premise was initially theorized by the Swiss mathematician Daniel Bernoulli in the 18th century. Bernoulli founded the idea regarding the differing values of things. That is, the ability of a good or service to satisfy "a" customer's needs or wants; with respect to theory, the utility of an item tends to be closely correlated to its price. An item such as gold, which is very useful and thus has great utility (combined with its scarcity), is very expensive. Total utility is closely tied to the bare concept of utility. Total utility points to the aggregate amount of usefulness and fruition there is to be gained from the use of a specific good, service, or other item. Furthermore, the abstract measurement of utility is another key concept of the theory. Although it is hard to calculate the exact utility of something, economists use abstract measurements to capture the usefulness of things. Graphically, utility is presented in Figure 5.1. Notice the axes (the y-axis is marginal utility and the x-axis is quantity consumed).

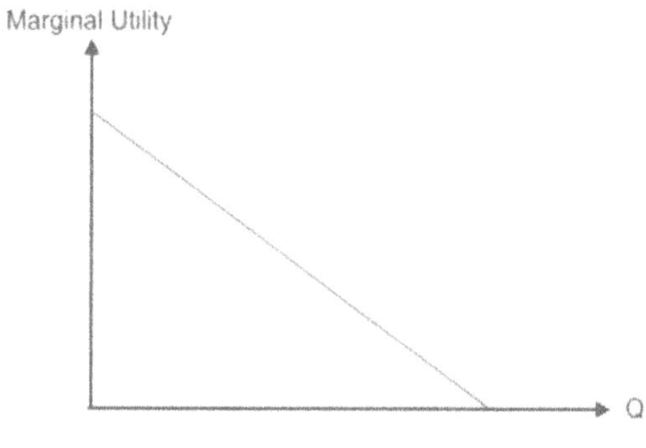

FIGURE 5.1 Marginal utility versus quantity consumed.

DOI: 10.1201/9781003650614-5

One may calculate the marginal utility, thusly: MU = change in total utility/change in quantity. The reader will notice that marginal utility decreases as consumption of a good increases. This illustrates the Law of Diminishing Marginal Utility.

There are many steps that businesses can take to improve utility for their customers. This includes research and marketing activities, such as focus groups and testing. Companies can also consider increasing the speed with which they conduct their production process, resulting in the ease of bringing products and services to market. Companies can also make their products and services easily available (in retail locations and online) at lower costs.

ALL this effort is because people, in the final analysis, purchase goods and services to get some benefit or satisfaction. This allows them to fulfill a need or want when they consume it. This phenomenon is called economic utility. There are six principles that fall under this umbrella, including form utility, time utility, place utility, information, service, and possession utility. Companies can boost their sales and revenues by understanding and tailoring their marketing and production efforts to the way individuals purchase and consume their products. For more on this topic, see J. B. Maverick at: https://www.investopedia.com/contributors/53889/ Updated December 4, 2023. Here, we will summarize the concept of economic utility thusly.

From a quality perspective, *Economic Utility* can be defined as the total amount of satisfaction that someone experiences when they consume a particular product or service. It helps measure how much fulfillment someone requires to satisfy a particular need or want. Generally, companies strive to increase the utility or perceived value of their products and services to enhance customer satisfaction, increase sales, and drive earnings. The concept of economic utility falls under the area of study known as *behavioral economics*, which is designed to assist companies in operating a business and marketing the company to attract the maximum number of customers and sales revenues.

As mentioned earlier, there are six types of economic utility, which include form, time, place, possession, service, and information. Companies that can understand and recognize areas that are lacking in their marketing schemes can assess consumer purchase decisions and pinpoint the drivers behind those decisions, thus boosting their sales and profits. So let us examine each of them.

Economic utility is the total amount of satisfaction experienced when a product or service is consumed. Economic utility can also be referred to as *utility marketing*. That is because product development and design require companies to persuade consumers to make purchases. So, what is an example of economic utility? The term economic utility refers to the total degree of satisfaction someone gets from using a product or service. It may be a car, house, food, clothing, financial services, or housekeeping. Companies that offer them can study the behaviors of their consumers and figure out what drives them to make purchases. One example of an economic utility is the value customers receive from the latest iPhone model. Apple responds to the needs and wants of its consumers by updating and upgrading its phones on a regular basis. So, what Are the six types of Economic Utility? Here we present them in a very concise form.

1. The first is *form utility*, which means the amount of value someone receives from goods or services that they "actually" need. *Form utility* is, therefore,

the incorporation of customer needs and wants into the features and benefits of the products being offered by the company.

Companies invest time and money into product research to pinpoint exactly what products or services consumers desire. From there, company executives strategize on the development of the product with the goal of meeting or exceeding those needs to create form utility. Form utility may include offering consumers lower prices, more convenience, or a wider selection of products. The goal of these efforts is to increase and maximize the perceived value of the products. For instance, a cosmetics company may conduct focus groups and testing to identify holes in the market related to different skin types and skin tones. The company may decide to produce and market new offerings to cater to and complement the needs of a more racially diverse clientele. The company can increase its sales while adding value to these new consumers.

Utility does not necessarily have to be measured in numbers – just in perceived value. For instance, someone may choose to walk rather than take the bus or drive because they perceive the health benefits from the exercise to be greater than the speed and ease associated with being transported in a vehicle.

2. *Time utility* has to do with the amount of time it takes for companies to respond to the needs and demands of their consumer bases. This type of utility occurs when a company provides goods and services when consumers demand or need them. Companies analyze how to create or maximize the *time utility* of their products and adjust their production process, logistical planning of manufacturing, and delivery. So, when demand increases, the company should respond by producing and delivering more of the product to the market.

   Creating time utility includes considering the hours and days of the week a company might choose to make its services available. For example, a store may open on weekends if customers typically shop for a certain product at that time. Time utility might also include 24-hour availability for a product or the company's customer service department through a phone number or website chat function. Failure to factor time utility into the equation can lead to a drop in the customer base, which can result in a loss of revenue.
3. The *third utility has to do with place,* which refers to a centralized location where consumers can easily access the products and services they need. *Place utility* involves making products or services available in locations that allow consumers to easily access them. Place utility refers to making goods or services available in locations that allow consumers to easily access products and services.

   While most people typically think of place utility as a physical or brick-and-mortar location, such as a retail store or shopping mall, the digital age helps broaden the definition of availability. For instance, companies can maximize place utility through their website. Those with effective search engine optimization strategies can improve their place utility. Increasing convenience for customers can be a key element in attracting business.

For example, a company that offers easy access to technical support gives consumers an added value compared to a similar company that does not offer a similar service.

Making a product available in a wide variety of stores and locations is considered an added value since it is more convenient. Apple sells iPhones and laptops through its retail stores, but also offers its products through other electronics retailers, including Best Buy.

4. *Possession utility* is yet another type of economic utility. It measures a product or service's perceived value based on a consumer's ability to obtain and use it as soon as the need or want arises. *Possession utility* is the use or perceived value a consumer gets from owning and being able to use a product or service in a timely manner. Possession utility is the amount of usefulness or perceived value a consumer derives from owning a specific product and being able to use it as soon as possible. The basic premise behind this utility is that consumers should be able to use a specific good or service as soon as they are able to purchase or obtain it. For instance, someone who purchases the latest iPhone will not get much utility for the product if Apple has it on backorder and cannot manufacture and ship it to the consumer in a timely fashion. That is why it is important for companies to increase the ease of ownership, which boosts the product's possession utility or perceived value. Consider lenders who offer favorable financing terms for owning a car, appliance, or home. They would likely create possession utility for these products, leading to an increase in sales and, therefore, revenue.
5. *Information utility*: the value given to a product by virtue of the fact that it can provide the user with information that is useful. *Information utility* refers to any data that helps consumers make buying decisions. This includes product details on e-commerce pages, targeted marketing campaigns, and well-trained call center and in-store agents capable of answering customer questions. Therefore, an excellent example of information utility is when a wholesaler provides customers with a sales catalog featuring the various products for sale. Additional information utility examples are targeted marketing campaigns, test drives, and product samples.
6. *Service utility*: Service utility involves providing customer support before, during, and after the purchase/service. *Service utility* is the utility created when services are rendered by professionals. For example, the services of doctors to their patients, the services of teachers to students, the services of lawyers to clients, *etc*. The utility creation is inseparable from the service provider. Here, both production of utility and its consumption take place at the same time.

With an introductory knowledge of economic utility, now let us focus on the perceived value of a product, service, or item. Perceived value comes down to the price the public is willing to pay for a good or service. Even a snap decision made in a store aisle involves an analysis of a product's ability to fulfill a need and provide satisfaction compared to other products under different brand names. See more on perceived value on https://www.investopedia.com/terms/p/perceived-value.asp, the article by J. Kopp.

The work of the marketing professional is to enhance the perceived value of the brand they are selling. The pricing of products takes perceived value into consideration. In some cases, the price of a product or service may have more to do with its emotional appeal than with the actual cost of production. Even a snap decision made in a store aisle involves an analysis of a product's perceived ability to fulfill a need and provide satisfaction.

In marketing terminology, perceived value is the customers' evaluation of the merits of a product or service, and its ability to meet their needs and expectations, especially in comparison with its peers. To be sure, marketing professionals try to influence consumers' perceived value of a product by describing the attributes that make it superior to the competition.

It is very important here to mention that the perceived value is a customer's own perception of a product or service's merit or desirability to them, especially in comparison to a competitor's product. Consequently, it is measured by the price the public is willing to pay for a good or service. The marketing of a product or service, therefore, involves attempting to influence and increase its perceived value, which can emphasize qualities such as its aesthetic design, accessibility, or convenience. So, marketers who want to influence the perceived value of a product define its attributes in terms of its utility, or the extra benefits and values that the customer expects to get in using it. The perceived utility of many products and services may differ widely even among similar or virtually identical products.

So, an effective marketing campaign for taking advantage of perceived value by the customer is the one that focuses on the following five items:

1. *Form utility* is the aesthetic appeal of the physical design of a product. Even a utilitarian product like a frying pan can increase in perceived value because of its appealing design.
2. *Task utility* is the value attached to a service that saves the customer time, effort, or money. Car detailing shops and laundry services offer utility value.
3. *Time utility* refers to the ease of access to a service or product, such as a 24-hour service compared to a 9-to-5 service.
4. *Place utility* is the convenience of the location, like a fast-food outlet that's around the corner compared to a restaurant that's 20 miles away.
5. *Possession utility* refers to the ease of purchasing the product. A department store that features online ordering, home delivery, or in-store pickup is aiming for possession utility.

SPECIAL CONSIDERATIONS OF PERCEIVED VALUE

A company's brand is meant to communicate a set of expectations associated with its products or services. That is why a well-established brand can command a higher price than its generic equivalents. Advil and Motrin both contain ibuprofen, but both brands are priced higher than generic ibuprofen.

Luxury goods, however, carry the perception of value to another level with the addition of prestige. The highest value of luxury goods is not associated with their utility but with the prestige that owning and using them entails. The perceived value of a Rolex watch is not based on its functionality but on its image as a mark of personal success and refined taste.

At the opposite end of the scale, some brands are marketed as smart bargains. The perceived value of a product may be its low price in comparison with competitors of equal quality. In the end, it does not matter if the product or service is not commensurable with the expected prestige. We see this especially in the auto industry. All major companies had record recalls in 2023. For the details, see B. Cline's article in: https://www.lemonlawlawyerscalifornia.com/2023/12/top-10-automakers-with-the-most-car-recalls-in-2023/#:~:

The implication in this chapter is obvious. There is a close relationship between economic utility, marketing and quality. In fact, they are all interlinked. We already made a few comments on the utility issue, so now let us extract the relationship between marketing and quality. Marketing focuses on capturing and analyzing customers' needs – through some kind of feedback from the customer; quality involves the question of whether products and services do correspond to customers' expectations and needs, and whether both are congruent with each other. Above all, however, they are consistent and on a timely basis.

A company may work hard to get the attention of buyers, but it is the quality of a product over time that sustains this attention and converts buyers to customers, which eventually leads to brand loyalty. Consistency in the quality of a product over time is what gives rise to brand loyalty.

Having a high-quality product can increase positive associations with your brand, including desirable qualities like consistency and an understanding of customers' needs. This, in turn, can strengthen brand recognition and loyalty, and increase the likelihood of long-term success.

When quality and marketing are coordinated, this is called customer-oriented approach, and its key priority is customer satisfaction. Its main objective is to produce products and services as per customers' desires and make them available to customers at the right place, at the right time, and at affordable prices.

To consistently exceed our customers' expectations, we believe quality is an integral part of the changes. And that means we need a quality strategy that details our test approach for developing and delivering a high-quality product to the customers. In essence, then, marketing serves as a bridge between consumers and the resources they need to lead fulfilling lives, ultimately contributing to the improvement of the overall quality of life. Furthermore, if we perpetuate a policy of poor data quality (in all forms) it can damage not only CS but also future marketing campaigns, which may lead to ineffective targeting, subpar personalization, wasted resources, and damaged brand reputation, as well as profits. However, there is a warning in this discussion: Be aware of a higher price because they may decrease consumer utility, since they must pay more for the product. On the other hand, higher price may induce higher quality perceptions, which increase utility (Monroe, K. and Krishnan, R. (1985). "The Effect of Price on Subjective Product Evaluations." In: Jacoby, J. and Olson, J., Eds., *Perceived Quality: How Consumers View Stores and Merchandise.* Lexington Books. Boston, MA, 209–232. See also: https://www.researchgate.net/post/How_can_marketing_improve_quality_of_life/65cc7a9aa96c5db09b0ee410/citation/download).

6 The Role of Management in Customer Service

OVERVIEW

In all industries from aerospace, automotive, defense, agricultural to medical devices, hospitality, health, to consumer-packaged goods, companies cannot afford to let defective products reach customers. While only a small percentage of recalls are typically related to production issues, scrap, rework, and warranty-related costs can total into the millions for an individual manufacturer, with reputational risk being unquantifiable. I am reminded of M. Hersey, who used to say (I am paraphrasing him): "Quality and CS is the best advertisement." What he meant was that quality, and specifically CS, must be consistent and continual.

So, for any organization, they are two primary roles for any executive in any organization as they "tackle" CS. They are:

1. **Top service priorities. They include at minimum:**
   a. Improvement of customer experience
   b. Improvement of data quality and reliability
   c. Improvement of workforce skills (either internal or external training)
2. **Top service Challenges. They include at minimum:**
   a. Making sure to keep up with customer expectations
   b. Making sure that the pressure to reduce cost to serve is continual
   c. Making sure that poor data quality/reliability are "always" replaced with current and informative ones

THE ROLE OF STRATEGIC LEADER

Leaders provide direction, vision, motivation, and inspire others (inspiration), and help create an environment conducive to success by promoting communication and collaboration among team members. In short, leadership and strong management are essential for any organization that wants to achieve its objectives. Therefore, quality is to any organization, equivalent to the ethics and character of an individual.

The direct responsibility of management is critical to customers today because more than ever, the customer expects products with zero defects and 100% on-time performance, yet faster rates of production development cycles and more varied product mixes introduce opportunities for errors – and rising risks around cost of quality.

Therefore, because of this importance and linking CS with organizational cost, it is imperative to use the classic approach of the 1-10-100 rule to push down the cost of quality. [The 1-10-100 Rule was born from both experience and research. In data quality, the cost of verifying a record as it is entered is $1 per record. The cost of

DOI: 10.1201/9781003650614-6

remediation to fix errors after they are created is $10 per record. The cost of inaction is $100 per record per year.] Of course, there are other tools as well. For example, the implementation of layered process audits may optimize the power of prevention and early detection to reduce costs. Another way may be to use digital transformation – there are many software packages in the market that can help in both identifying unnecessary costs and quality trends at the same time.

In any case to optimize the CS, any management leader should have objectives for tackling current problems as well as preventing problems. Typical objectives should cover at least the following:

- Develop their leadership capability through self-evaluation, development planning, practice, and application.
- Increase their ability to lead change through an engaged team.
- Develop team capability through effective feedback, coaching, empowerment (give subordinates responsibility and authority to act), and development.
- Lead, inspire, and motivate others through effective conversations.
- Increase their ability to influence others and leverage networks in a matrix environment.
- Deepen their cultural understanding to work more effectively across the global enterprise.
- Increase their strategic orientation and their cross-functional connection to specific organizational strategy.
- Stimulate their innovation mindset.
- Learn how to lead high-performing teams.

Lofty objectives? Maybe. However, our world today is changing exponentially, with new technologies being developed at a rate that we (society) can hardly keep up with. Therefore, in today's world, we need a new way of thinking about technology and our future survival. This means that we must be ready to expect emergence over authority more often than in the past. Specifically, it means: (1) The recognition that *Not knowing* is okay. In fact, we have entered an age where the admission of ignorance offers strategic advantages over expending resources –subcommittees and think tanks and sales forecasts – toward the increasingly futile goal of forecasting future events. (2) Recognition of *asymmetry* because effects are no longer proportional to the size of their source. Nowadays, it is not unusual for the small players to be the main change makers. (This is very common in the health, pharma, and energy sectors.). (3) Complexity in our modern environments is defined quite often as heterogeneity, interconnection, interdependency, and adaptation. So, when these three conditions are in place, certain guiding principles serve us best. These are the Uniqueness, Impact, and Magic elements. *Uniqueness* means that if someone is working on a similar project elsewhere, the organization should move or adapt to something else. Here we must emphasize something that most of us have been told many times: Knowledge is power. However, knowledge on its own – for its own sake – even though beneficial, in the real world of organizations, the gained new knowledge must have an *impact* on either the start-up or physical creation of a product or service. If not, it has created busy work for conversation. On the other hand, the *Magic* element, though hard to define, speaks to the delight that playful originality so often awakens. In the final analysis, we really do not have a choice of not adopting the age of acceleration, if we want to be competitive and please the customer. If we do not, the competitors will. For a more detailed exposition on these thoughts, the reader is encouraged to see *Thank You for Being Late: An Optimist's Guide to Thriving in the Age of Accelerations*, by Thomas Friedman, 2016. Farrar, Straus, and Giroux Publishers. NY, NY.

It may be difficult, costly, and time-consuming but with appropriate training, commitment, and enthusiasm for growth, *one may start by a personal inventory* of the following:

1. What do I say
   a. How I frame the issues
   b. What I emphasize
   c. What I repeat
   d. Context content
2. How I act
   a. Relationships
   b. behaviors
   c. Symbols
3. My priorities
   a. My standing meeting
   b. Who I interact with
   c. Disciplines
   d. Routines
4. What I measure
   a. Rewards
   b. Recognition

In order for the change to occur, we must come to a realization that the management leadership must practice strategic approaches for the growth and profitability of the organization. The role of the strategic leader's responsibility is shown in Table 6.1.

TABLE 6.1
The Role of the Strategic Leader

From an Individual to Team Focus	The Strategic Leader
Focus on team goals and implementation of strategy. The aim is success.	Focus on one (no more than five at the time) organizational goals and the implementation of the organizational strategy for success.

The Flow

→ Personal, Team, Organization (Skill teams), Enterprise (working across skill teams)

So, the transformation of the leader is to start working strategically. Working strategically means: Strategic Thinking + Planning + Execution.

1. *Strategic thinking*: A broad, multi-directional mindset and awareness of the complete picture, proactively connecting the pieces to deliver desired business results.
2. *Strategic planning*: Doing the work to define each piece of the picture and the relationships between them, resulting in a clear, aligned, and actionable strategic plan.
3. *Strategic execution*: Marshaling/guiding people and resources toward achievement of the strategic plan, and facilitating governance and monitoring mechanisms to assess progress and ensure accountability.

A typical roadmap for success may look something like Table 6.2.

To verbalize this success in a summary form, the following characteristics must take place without any circumvention of avoidance:

1. Organizational strategy
   - Plan
   - Actual status against plan
   - Forecasted status against plan
   - Risks and opportunities
   - Better plan
   - Priority areas of focus and key metrics
   - Performance
   - Customers and markets
   - Products and services
   - Processes and people
2. External/internal environment assessment
   - Strategic business challenges and opportunities
   - Business environment and opportunities
   - 2 Vision (if revised)
   - Mission
   - Vision
   - Values/behaviors
   - Customers and markets
   - Products and services
3. Strategy (priorities)
   - Priority areas of focus and key metrics
   - Performance
   - Customers and markets
   - Products and services
   - Processes and people

TABLE 6.2
Roadmap to Success

An Overview of a Roadmap to Success

Must have:	Create an:	Generate:
- Vision and strategy - Personal leadership vision [Vision: People working together as a lean, global enterprise for the organization's goals]	Organizational strategy [A typical strategy may be: 1. Analyze the business environment: a. External/internal environment b. Strengths / weaknesses / opportunities / threats c. Benchmark organization attributes 2. Clarify / create the organization's strategic intent (vision): a. Core customers and markets b. Core products, processes, and services c. Vision statement d. Mission statement 3. Establish strategic priorities and metrics: a. Scope – 3 to 5 years, organization level (not department) b. Number – no more than six areas of focus c. Impact – address/leverage business environment and strategic intent work d. Values and behaviors 4. Assign champions to drive the work and engage all employees in the plan (plan objectives): a. Definitions/milestones/metrics b. Team objectives c. Individual objectives d. Project plan e. Deliverables 5. Deliver results through strategic execution a. Governance to monitor plan – Business Plan Reviews (BPRs) b. Special Attention Reviews (SARs) c. Project plans]	ONE plan and goal at a time [Typical outputs for a plan, are: 1. Outputs a. Environment analysis b. Vision and mission c. Strategic priorities and metrics d. Objectives and targets e. Governance]

TABLE 6.3
Final Plan

Now (*Yesterday to today*)	Strategic Priorities	Action Steps (*Now to the destination*)	Destination (*Our vision and destination*)
Current state: Where are we?	Key areas to focus on: Where should we focus?	Actions required to get us there: How will we get there?	Vision and direction: Where do we need to go?

4. Plan (objectives)
    - Priority areas of focus and key metrics
    - Performance
    - Customers and markets
    - Products and services
    - Processes and people
5. Business plan and review (Final is shown in Table 6.3.)
    - Plan
    - Actual status against plan
    - Forecasted status against plan
    - Risks and opportunities
    - Better plan

So, the whole idea of the leadership's commitment to CS is, in fact, innovation with the intent to be better. That is supported by the statistics given in the previous chapters and many more in the literature for improved CS. As important as innovation is, however, leadership must understand that the system of innovation must in some way be standardized. There must be a system that will generate a flow of innovations throughout the organization, and it must be encouraged for everyone to participate.

Although the word standardization may seem contradictory to CS, the fact is that we have many systems within quality where standardization takes place. For example, the ISO 9000 standardizes the Quality System of all organizations, with worldwide improvements published in many publications. There is also a specialized standard for innovation. The ISO 56001 is such a standard focused on innovation management. It focuses on the general structure for organizations to develop, implement, and improve innovation processes. That means the standard's intent is to help enhance the creation and delivery of value through new ideas, products, or services for any organization. A warning here is warranted. Do not over-standardize. It is obvious, but if you do, you will limit breakthrough ideas, slow the drive for progress, and restrict creativity.

The results of this "intent" are to streamline processes, reduce costs, and boost collaboration. Obviously, as part of innovation, creativity is a major part of this endeavor, without minimizing consistency and encouragement of active participation by all.

Three other specific tools that may be used in implementing innovation in CS are TRIZ, lean manufacturing, and benchmarking. TRIZ equips organizations with systematic tools for inventive problem solving, while lean principles emphasize efficiency and waste reduction. Benchmarking allows the organization to use "best practices" in each industry, or how they compare to your own. Together, they can create a dynamic environment for creativity to flourish alongside operational excellence.

And so, now the leadership focuses on whether or not to accept innovation management principles. To make that decision, they must address the following issues for the entire organization. They are:

1. *Organizational readiness*: Two serious questions must be answered. Is the organization ready for the change? Are the resources available for the change?
2. *Potential risks*: Are the risks commensurable with the anticipated benefits? Is mitigation available for alternatives?
3. *Compliance and external expectations*: Are you doing the innovation strategy because of your competitors' pressure or government regulations? Or, are you doing it to stay ahead of the competition?
4. *Strategic alignment*: Changes must be able to align the goals of the organization with customer satisfaction. For example: Costs, personnel available for the proposed changes? Do the changes meet the expectations of the organization's ROI goals?
5. *Performance metrics*: Are the new or potential Key Process Inputs (KPI) clear and measurable to evaluate correctly the innovation efforts?
6. *Customer and market effect*: Are the changes better for the customer, and will these changes help us be ahead of the competition?
7. *Stability and adaptability*: How flexible is your organization? Can it support incremental and disruptive innovation?
8. *Sustainability*: Are your innovation recommendations compatible with your organizational goals? Are they long-term and success-oriented?
9. *Employee and culture engagement*: Perhaps the most important responsibility of management's role in implementing the innovation strategy is employee and culture engagement. A proper evaluation is necessary. Without this evaluation, your strategy will fail. On the other hand, appropriate gauging and measurable assessment of the employee and culture will facilitate: Engagement, collaborative, and above all an innovative workplace culture.

7 Problem-Solving Methodology

OVERVIEW

Perhaps the most misunderstood area in any organization is the understanding and selection of a problem-solving methodology. It is misunderstood, maybe because there are so many options to choose from, and depending on whom you ask, the answer to the best approach is quite different. However, the fact of the matter is that all organizations in the modern world must have a method – preferably standardized – across their facilities, whether local, national, or international.

So, what is a problem-solving methodology? Problem-solving methodology is a strategy with specific steps that one would use to find the problem(s) that are in the way of accomplishing one's own goal. Another way of saying is that problem solving consists of using generic or *ad hoc* methods, in an orderly manner, for finding solutions to problems. Others, of course, may have different definitions, for example, calling it as the problem-solving cycle (PSC) (Bransford and Stein, 1993). In this cycle, one will recognize the problem, define the problem, develop a strategy to fix the problem, organize the knowledge of the problem cycle, figure out the resources at the user's disposal, monitor one's progress, and evaluate the solution for accuracy. Although called a cycle, one does not have to do each step in sequence to fix the problem. In fact, in some cases, if the problem is simple, not all steps may be necessary. However, the reason it is called a cycle is that once one is completed with a problem, another usually will surface. [No matter how you define the problem, in the final analysis, it is important to recognize that all problems fall into three categories: (1) Products or processes are the result of degrading over time from the acceptable performance, (2) abrupt change from the *status quo*, and (3) problem is the result of failing to meet an unusual goal. That is, the goal is beyond the process' capability.]

Blanchard-Fields (2007) looks at problem solving from one of two facets. The first is looking at those problems that only have one solution (like math/engineering problems, or fact-based questions), which are grounded in psychometric intelligence. The second is addressing socio-emotional issues, which are unpredictable and yield answers that change over time (like what is your favorite color or what you should get someone for Christmas). To show the diversity of problem-solving strategies, we provide the reader with a sample of techniques that are available. By no means are these an exhaustive list.

- *Abstraction*: Solving the problem in a model of the system before applying it to the real system.
- *Analogy*: Using a solution that solves an analogous problem.

DOI: 10.1201/9781003650614-7

- *Brainstorming* (especially among groups of people): Suggesting many solutions or ideas, and combining and developing them until an optimum solution is found.
- *Divide and conquer*: Breaking down a large, complex problem into smaller, solvable problems.
- *Hypothesis testing*: Assuming a possible explanation to the problem and trying to prove (or, in some contexts, disprove) the assumption.
- *Lateral thinking*: Approaching solutions indirectly and creatively.
- *Means-ends analysis*: Choosing an action at each step to move closer to the goal.
- *Method of focal objects*: Synthesizing seemingly non-matching characteristics of different objects into something new.
- *Morphological analysis*: Assessing the output and interactions of an entire system.
- *Proof*: Try to prove that the problem cannot be solved. The point where the proof fails will be the starting point for solving it. Some call this method the *escape point*.
- *Reduction*: Transforming the problem into another problem for which solutions exist.
- *Research*: Employing existing ideas or adapting existing solutions to similar problems.
- *Root cause analysis*: Identifying the cause of a problem.
- *Trial-and-error*: Testing possible solutions until the right one is found.

So, even though there are many strategies to approach a problem, one must recognize that no matter what the strategy is, all of them have advantages and disadvantages. Therefore, the organization must select a methodology that fits its culture and processes. Some common ones are:

1. *Applied problem solving (APS)*: The model is based on the classic approach of solving problems, which are as follows:
   a. Comprehension
   b. Representation
   c. Planning, analysis, and synthesis
   d. Execution and communication
   e. Evaluation
      i. *Comprehension*: A model where making sense of the problem is by using strategies such as retelling, identifying relevant information, and creating mental images. This can be helped by encouraging participants to re-read the problem several times and record in some way what they understand the problem to be about (for example, by drawing a picture or making notes).
      ii. *Representation*: A model of "homing in" on what the problem is, and asking solvers to investigate. Central to this stage is identifying what is unknown and what needs finding. Typical questions may focus on:
         1. Can they represent the situation mathematically? (*i.e.*, Y = f (x, n); where Y = depended variable, x = independent variable(s), n = noise).

    2. What is it that they are trying to find?
    3. What do they think the answer might be (conjecturing and hypothesizing)?
    4. What might they need to find out before they can get started?
  iii. *Planning, analysis, and synthesis*: A model about planning a pathway to the solution, assuming that the participants have understood what the problem is about and established what needs finding. It is within this process that you might encourage participants to think about whether they have seen something similar before and what strategies they adopted then. This will help them to identify appropriate methods and tools. Particular knowledge and skills gaps that need addressing may become evident at this stage.
  iv. *Execution and communication*: A model where participants might identify further related problems they wish to investigate. They will need to consider how they will keep track of what they have done and how they will communicate their findings. This will lead on to interpreting results and drawing conclusions.
  v. *Evaluation*: A model where participants can learn as much from reflecting on and evaluating what they have done as they can from the process of solving the problem itself. During this model, participants should be expected to reflect on the effectiveness of their approach as well as other people's approaches, justify their conclusions, and assess their own learning. Evaluation may also lead to thinking about other questions that could now be investigated

2. *Goal, reality, obstacles/options, and way forward (GROW model)*. The particular value of GROW is that it provides an effective, structured methodology that helps set goals effectively and is a problem-solving process. G = Goal = This is the end point, where the organization wants to be. The goal must be defined in such a way that it is very clear to the organization when they have achieved it; R = Reality = The current reality is where the organization is now. What are the issues, the challenges? How far are they away from their goal? O = Obstacles = they are the items that the organization wants to remove. If there were no obstacles, the organization would already have reached its goal. O = Options = Once obstacles have been identified, the organization needs to find ways of dealing with them if they are to make progress. These are the Options. W = Way forward = The options then need to be converted into action steps, which will take the organization to its goal. These are the Way Forward.
3. *How to solve it*: A very useful approach is the mathematical approach to problem solving that focuses on four steps developed by Polya (1945, pp. 6–15). They are:
    a. First, you must understand the problem.
    b. After understanding, then make a plan.
    c. Carry out the plan.
    d. Look back on your work. Ask how it could be better?

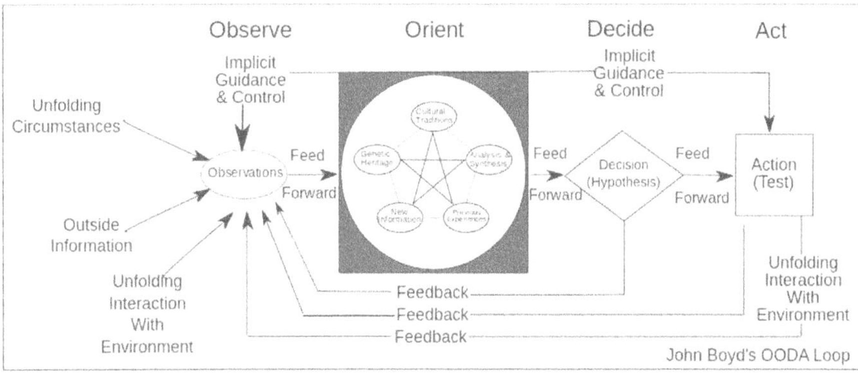

FIGURE 7.1 John Boyd's OODA loop.

   e. If this technique fails, Pólya advises: "If you can't solve a problem, then there is an easier problem you can solve: find it."

4. *Kepner-Tregoe problem solving and decision-making*: A model (some call it a technique) that refers to a process of weighing alternatives in which a person lists and assigns a numerical weight to a series of values (noting that some may be absolute requirements), gives each alternative a numerical rating according to each value, and computes a numerical score for the alternative (as the dot product). The technique allows that an absurd result may indicate an error in the weight and the rating, which to be solved by adjusting and iterating. A variation of this is the IS and IS NOT approach.
5. *OODA loop*: This is the observe, orient, decide, and act model designed by J. Boyd. The actual model, according to http://www.danford.net/boyd/essence.htm, is shown in Figure 7.1. See also Boyd, J. (1996). *The Essence of Winning and Losing*. p.3. http://stevenshack.com/johnboyd/assets/ essence of winning losing.pdf. See also https://en.wikipedia.org/wiki/OODA_loop. The model is in the open domain and may be used as needed. Also, a simpler version may be found on the website of Psych Safety.

   According to Boyd, decision-making occurs in a recurring cycle of observe-orient-decide-act. An entity (whether an individual or an organization) that can process this cycle quickly, observing and reacting to unfolding events more rapidly than an opponent, can thereby "get inside" the opponent's decision cycle and gain the advantage. Osinga (2006), however, argues that Boyd's own views on the OODA loop are much deeper, richer, and more comprehensive than the common interpretation of the "rapid OODA loop" idea.

   Boyd developed the concept to explain how to direct one's energies to defeat an adversary and survive. Boyd emphasized that "the loop" is actually a set of interacting loops that are to be kept in continuous operation during combat. He also indicated that the phase of the battle has an important bearing on the ideal allocation of one's energies.

Boyd's diagram shows that all decisions are based on observations of the evolving situation tempered with implicit filtering of the problem being addressed. These observations are the raw information on which decisions and actions are based. The observed information must be processed to orient it for further making a decision. This model has become an important concept in litigation, business, and military strategy.

Whereas the OODA loop focuses on strategic approaches and has been adopted by business and public sector entities, the difference with the PDCA or Shewhart cycle is that the PDCA model focuses on the operational or tactical level of projects.

6. *PDCA (Plan-Do-Check-Act) model*: This is the classic model of improvement used by many organizations worldwide. It was modified by Dr. Deming as the PDSA plan, do, study, and act. The difference between the two is that in using the "check" characteristic as opposed to "study" is the focus. With a check, the emphasis is on controlling by inspecting, whereas a study implies planning and prevention.
7. *RPR (rapid problem resolution)*: A specific model for identifying diagnostically the root cause of IT problems (Offord 2011). Specifically, it deals with failures, incorrect output, and performance issues, and its particular strengths are in the diagnosis of ongoing and recurring grey problems [a term for an IT problem where the causing technology is unknown or unconfirmed]. Common grey problems are: (a) Intermittent errors, (b) Intermittent incorrect output, or (c) Transient performance problems. The model is made up of two approaches. They are:
   I. *Core process*: They define a step-by-step approach to problem diagnosis and has three phases:
      i. Discover
         a. Gather and review existing information.
         b. Reach an agreed understanding.
      ii. Investigate
         a. Create and execute a diagnostic data capture plan.
         b. Analyze the results and iterate if necessary.
         c. Identify root cause.
      iii. Fix
         a. Translate diagnostic data.
         b. Determine and implement fix.
         c. Confirm root cause addressed.
   II. *Supporting techniques*: The supporting techniques detail how the objectives of the core-process steps are achieved, and cite examples using tools and techniques that are available in every business.
8. *TRIZ (in Russian: Teoriya Resheniya Izobretatelskikh Zadatch; in English: "theory of solving inventor's problems")*: it is a problem-solving analysis and forecasting tool derived from the study of patterns of invention in the global patent literature (Hua, et al 2006; Barry, et al 2010; Sheng and Kok-Soo 2010).
9. *8D*: A model that is used heavily in the automotive industry and has gained favor in many other industries either directly or in a modified version.

It is a model used to approach and resolve problems, typically employed by quality engineers or other professionals. Its purpose is to identify, correct, and eliminate recurring problems, and it is useful in product and process improvement. It establishes a permanent corrective action based on statistical analysis of the problem (when appropriate) and focuses on the origin of the problem by determining its root cause as well as the escape point of the problem. Although originally it was comprised of eight stages, or "disciplines," it was later augmented by an initial planning stage. 8D was developed by Ford Motor Company, and it is used widely all over the world in many manufacturing and non-manufacturing organizations. The augmented version in Ford Motor Company is called Global 8D (G8D), but other users continue to call it 8D. Both the 8D and G8D follow the logic of the PDCA model. A detailed discussion will be given in Chapter 8.

10. *5 Whys*: The 5 Whys is an iterative question-asking model/technique used to explore the cause-and-effect relationships underlying a particular problem. The primary goal of the model/technique is to determine the root cause of a defect or a problem. It is equivalent to the "pealing the onion" approach. The assumption is that if you keep asking questions, eventually you will get to a question which you do not know the answer. At that point, you declare it as the problem. After the definition of the problem, you pursue the analysis, and hopefully, the right decision will be reached.
11. *3×5 Whys*: A variation of the 5 whys. The difference between the two is that the 3×5 whys addresses issues on both the technical and system levels. A typical form is shown in Table 7.1.
12. *Lean methodology*: It is a method that focuses on waste reduction by using many tools, but the essence is based on three approaches to reducing any type of waste in a given organization. The three approaches are:
    a. *Stabilize*: To establish stability, excess movement of work and variation of processes is removed.
    b. *Standardize*: Once the stability has been reached, then formal standards via work rules, charts, and visual controls must be used to further reduce variation.
    c. *Simplify*: Simplicity means leveling the workloads among staff to ensure effective and efficient service is provided to the customer.

DEFINITION OF A PROBLEM

There are many ways to define what a problem is. However, here we use five of the traditional generic definitions.

- A problem is a SYMPTOM, not a cause or a solution.
- A problem is a gap between what is happening and what you want to happen.
- A problem is something that keeps you from achieving your goal.
- A problem could be a mistake, a defect, an error, a missed deadline, a lost opportunity, a non-conformance, or an inefficient operation. In the final

TABLE 7.1
A Typical 3×5 Why Form

Problem statement								
What was wrong?								
Where was it detected?								
When was it detected?								
Who was impacted?								
Non-conformance Problem statement	Why 1 Why the problem occurred?	Why 2	Why 3	Why 4	Why 5	Why 6 Technical root cause	Why 7 System root cause	Corrective Action
	Therefore	Therefore	Therefore	Therefore	Therefore	Therefore		
Non detection Problem statement	Why 1 Why was the problem undetected	Why 2	Why 3	Why 4	Why 5 Technical root cause	Why 6 System root cause	Corrective Action	
	Therefore	Therefore	Therefore	Therefore	Therefore			

analysis, a problem may be thought of as an OPPORTUNITY to do better. That is why we must always keep in mind that everything starts with people (their understanding and way of execution). That is, people have ideas, and those ideas are articulated in the work environment with the infusion of technology, hardware, or something else. The progression is always the same, the order never changes!

The idea here is that by defining the problem correctly in your problem-solving methodology, it will lead you to improvement. That improvement may be materialized through *Kaizen* and or *Breakthroughs*: *Kaizen* is a gradual, relentless, incremental process improvement over time, while *Breakthrough* is a sudden, significant improvement in process performance achieved in a very short interval of time (re-engineering). Obviously, continual improvement is best achieved by a combination of both Kaizen and Breakthroughs. Ideally, an organization should employ and encourage both Kaizen and Breakthrough in its improvement strategy; however, Kaizen is much easier to accomplish.

To optimize that improvement, one must recognize that there are two types of problems: The *sporadic* and the *chronic*. A sporadic problem is a sudden, elevated departure of the level of a problem from its historic, *status quo* level. On the other hand, a chronic problem is the difference between the historic level and an optimum or zero level; a chronic problem is a "chronic disease," which can economically be eliminated. A chronic problem is something that keeps on repeating over time. In order for both to be eliminated, management must have the appropriate attitude. This means that problems: (a) must not be viewed as burdens (b) must be recognized and dealt with, rather than ignored or pushed down the stream of processes, and (c) must be accepted as challenges or opportunities to improve.

These three attitudes are controlled by management, and as such, it is management's responsibility to:

- Create a motivating environment conducive to effective, companywide problem-solving methodology.
- Create a real opportunity for problem solving.
- Create a systematic approach (roadmap) for effective problem solving.
- Create the applicable knowledge of the tools for problem solving – via education and or training.

In addition to these requirements, there are also six essential ingredients that need to be in place for positive results. They are:

- Awareness of the importance of eliminating errors and the cost of errors to the business.
- Desire to eliminate errors.
- Training in proven, effective methods for solving and preventing problems.
- Failure analysis to identify and correct the real root causes of problems.
- Follow-up on tracking problems and action commitments.
- Recognition, giving liberal credit to all those who participate.

Unfortunately, many organizations talk positive about their problem-solving methodology, but when they discuss implementation procedures, they find many roadblocks (excuses) for not implementing an effective methodology. Some of the key roadblocks are:

- Lack of time.
- Lack of appropriate authority, responsibility, and accountability.
- Lack of problem ownership (if everybody is responsible, no one is responsible).
- Lack of recognition.
- Errors have been accepted as a way of life (this is how we do it here).
- Ignorance of the importance of a problem.
- Ignorance of the opportunity cost of a problem (potential savings).
- Belief that no one can do anything about some problems.
- Poor balance by upper management among cost, schedule, and quality.
- People who try to protect themselves (CYA) (fear of intimidation and ridicule).
- Head-hunting by management (selecting team members that are not competent for the task at hand, and or are not willing to pursue real improvement).

PROBLEM-SOLVING CYCLE

Any problem-solving methodology is a cycle, for it repeats itself even though the problems may differ. Why then use a structured problem-solving approach? A structured, systematic, problem-solving approach helps us to:

- Illustrate the relative importance of problems.
- Show the real root causes of problems.
- Guide problem solvers to get unstuck, know where they are at, keep on track, and persevere to results.
- Keep problems solved.
- Establish accountability and a motivating environment for problem solving.
- Produce consistently better solutions to problems than unstructured approaches.

As mentioned earlier, there are many options that an organization may follow to optimize its improvement and remove defects from its processes. In fact, successful, continuously improving companies around the world use structured problem-solving systems. Some specific approaches in use in the USA are shown in Table 7.2.

Fundamentally, all of them use a variation of the basic steps for any problem solving. They are:

- Select the problem (the symptom that causes concern(s)).
- Find the root cause.
- Correct the problem.
- Verify correction.
- Prevent recurrence, then recycle.

TABLE 7.2
A Comparison of Several Problem-Solving Methodologies

Xerox's 6 Step	GM's 4 Step	Daimler Chrysler's 7 Steps	DoD's 6 Steps	Classical 7-Step Approach	Ford's 9 Steps
1. Identify and select the problem. 2. Analyze the problem. 3. Generate potential solutions. 4. Select and plan a specific solution. 5. Implement that solution. 6. Evaluate the solution.	1. Identify the problem. 2. Analyze the problem. 3. Identify decision criteria. 4. Implement **** 5. Evaluate and verifying pending solutions. 6. Adoption 7. Deciding and implementing final solutions. GM also uses the 7-diamond approach as a preliminary process for immediate reaction to internal Quality issues. The first 4 steps are used to quickly determine if an out-of-standard condition (special cause) exists. The last three focus on whether or not Engineering support is required.	1. *Problem*: Poor logo quality. 2. *Containment*: Verify process and final product. Quality alert. 3. *Root cause*: Machine setup, old equipment-> needs adjustment. 4. *Action*: Give maintenance, modify preventive maintenance plan. 5. *Verification*: 6. *Control*: 7. *Prevention*: Apply to similar processes/ products, modify FMEA.	1. Identify the problem. 2. Gather facts/make assumptions. 3. Define objectives, goals 4. Develop best solutions. 5. Analyze, compare solutions. 6. Select and implement solutions.	1. Identify the problem. 2. Contain the problem. 3. Define root cause. 4. Implement permanent action 5. Verify action success. 6. Implement process controls. 7. Implement preventive measures.	See chapter 8

These steps – see Table 7.2 – are based on three fundamental premises: (a) they are easy to understand and easy to remember (b) they are designed to be completed in sequence, and (c) there are also designed to have built in several iterations or feedback loops that may have to be followed to properly complete the cycle (task). Since they are important steps, let us examine each of them separately:

Step 1 - Select the problem (or opportunity) that represents a waste or a bottleneck in the area. Applicable actions may be to:

- Prepare a problem list.
- Collect data to determine the magnitude of each problem.
- Prioritize the problems to determine which problems should be worked on first. Almost always, a "vital few" problems account for the greatest proportion of the losses.
- Select the target problem, generally the problem whose solution will have the greatest beneficial impact.
- Prepare a specific problem statement (*e.g.*, Reduce shipping errors from 3.1% to under 1.5% in the next 12 months). This represents the specific problem assignment or project committed to. A complete problem must have three components: (a) the item (b) measurement, and (c) criteria in order to be complete. In the example of the reduction of errors, we have: the item of concern is shipping errors; the measurement is from 3.1% to 1.5%; and, criteria are the next 12 months.

Step 2 - Find the root cause of the problem. This analysis step is usually the crucial work of the project. Problems must be solved at the root cause level and must be doable within a reasonable time and cost, if they are to remain fully and permanently corrected. Applicable actions may be to:

- Identify all the possible causes of the problem.
- If necessary, collect data to ascertain the failure mechanism.
- Select the most likely cause. [Remember: there are three options here: (a) probable cause, possible, and likely cause.] Generally, a root cause cannot be fixed unless we find the escape point, *i.e.*, the place that we could have caught the problem, but we did not. The escape point indicates a system failure.

Step 3 - Correct the problem, following a plan that will eliminate the problem or at least reduce it to a level compatible with the team's goal. Applicable actions may be to:

- Make a temporary fix, if appropriate, to protect the customer until a permanent fix can be implemented.
- Develop alternative solutions for correcting the problem.
- Select the best possible solution by narrowing down the alternatives using a priority setting approach and making a final, consensus decision.
- Develop an implementation action plan, which includes a time schedule for implementation. Everyone who is involved in carrying out the solution should be consulted and should approve the plan. The plan should answer the basic questions: what? why? where? when? and how? Avoid the who? (IF you do not, the problem will focus on personalities and individuals, and thereby it becomes confrontational.)

- Establish the method for measuring the success of the proposed solution. Remember, the criterion for success was established as part of the problem statement in step 1.
- Gain management approval, if necessary, for implementation.
- Implement the action plan. [Assign a responsible person for the action plan and make sure there is a date of completion for the task. If there is no accountability and no responsibility, nothing will get done.]

Step 4 - Verify correction, using the criterion for success established in step 1 and the measuring method established in step 3. Applicable actions may be to:

- Measure the impact of the solution.
- Make and decide on "a" decision: Is the problem solved to a level compatible with the team's goal? Have any unforeseen new problems been created by the solution?
- If the correction is unsatisfactory, the team must go back to an earlier step, selecting an alternative solution or possibly even selecting an alternative root cause, then proceeding.
- If the correction is satisfactory, any temporary protective fix should be removed.

Step 5 - Prevent recurrence of the problem. This is an often-overlooked aspect of problem solving, the benefits of which cannot be overstated. Applicable actions may be to:

- Develop innovative methods for mistake-proofing the system (*i.e.*, Japanese "Poka yoke").
- Modify systems, FMEAs, instructions, and procedures to reflect the new, optimal practices.
- Train all involved personnel in the new methods.
- Publicize the results. Apply the knowledge gained to the rest of the product line and to other company activities with similar conditions.
- Continue to monitor the problem.
- Recognize the team's success.
- Prepare a final project report which describes the problem, the method(s) used to correct it, the quality, and productivity gains achieved. Also, a final team report might be made to management.

Recycle. Once the above 5 steps have been completed, the team should return to step 1, select another significant problem for correction, and continue improvement efforts.

COMMON TOOLS FOR PROBLEM SOLVING

In any problem-solving process, teams need to learn and apply a number of quality tools, *i.e.*, analytical techniques, to aid them in their working through the 5 steps of the PSC. These tools are primarily used for collecting, analyzing, and understanding data or information. To be sure, there are numerous quality tools that may be applied to problem solving. Some of these are very complex, advanced mathematical

TABLE 7.3
An Inventory List of 38 of the Most Commonly-Used Tools

Quality Tools Inventory	Basic 7	Selection	Cause	Corr	Verif	Prev	Data
ANOVA		X	X			X	V
Audit		X			X		A
Brainstorming		X	X	X			A
Cause and Effect	*	X	X				A
Check sheet	*	X	X		X		A
Control Charts (1)	*	X	X		X	X	VA
Customer Complaint Log and tracking		X			X	X	A
Customer Feedback Survey		X			X	X	A
Design of Experiments			X				VA
Evolutionary Operation (EVOP)			X				V
Failure Mode & Effects Analysis (FMEA)		X				X	A
Failure Analysis		X	X				A
Flow Chart		X	X				A
Force Field Analysis	*	X					A
Goal Setting		X					A
Group Vote		X	X	X			VA
Histogram	*		X		X		V
Key Characteristic Selection			X				A
Measurement System Evaluation (MSE)		X	X				V
Mind Map		X					A
Mistake Proofing		X				X	–
Nominal Group Technique		X	X	X			A
Pareto Chart	*	X	X		X		A
Pilot Run					X		–
Process Capability Study		X	X		X		VA
		X					
Process Log/Events Log		X	X		X	X	A
Process Potential Study		X	X		X		VA
Process Standardization and or Qualification						X	–
Cost of Quality Analysis		X			X		V
Quality Function Deployment (QFD)		X				X	–
Regression Analysis			X				V
Response Surface Methodology (RSM)			X				V
Scatter Plot	*		X				V
Statistical Process Control (SPC)		X	X	X	X	X	VA
Stratification			X		X		VA

(Continued)

TABLE 7.3 (*Continued*)
An Inventory List of 38 of the Most Commonly-Used Tools

Quality Tools Inventory	Basic 7	Selection	Cause	Corr	Verif	Prev	Data
Taguchi Methods			X	X			V
Team Meeting Guidelines		X	X			X	–
Test of Hypothesis			X				VA

V = Variable Data; A = Attribute Data

Control Charts = Variable Charts, Attribute Charts, (run charts, multi-Vari Charts, Box Plots, Pre-control, and so on)

For each tool listed, each step of the problem-solving cycle in which that tool can be utilized is indicated. For example, the Check Sheet may be used in Step 1 - Problem selection, Step 2 – Root-cause analysis, and Step 4 - Verification of correction.

These tools (1) can be easily learned and used by everyone in the organization, (2) are very effective in achieving basic problem-solving success, and (3) are essential to any properly designed improvement strategy.

Basic instructions describing these tools and their use may be found in any SPC book and or basic statistics book, as well as in the Memory Jogger.

TABLE 7.4
The 7 Basic Tools

Seven Basic Tools of Quality	Seven Basic Tools of Management
1. Check sheet (tally sheet)	1. Affinity diagram
2. Cause and effect diagram (fishbone or Ishikawa diagram)	2. Relationship diagram
3. Stratification.	3. Tree diagram
4. Histogram.	4. Matrix chart
5. Pareto chart (80-20 rule)	5. Matrix data analysis chart
6. Scatter diagram.	6. Arrow diagram
7. Control chart (Shewhart chart)	7. Process decision program chart

techniques, while others are quite simple to learn and apply. Table 7.3 shows an inventory list of 38 of the most commonly used tools. Appendix B also has many specific tools and their uses in any problem-solving endeavor.

If the problem-solving process demands an understanding of relationships, then the focus is on a slightly different approach to find the following: (a) what is the problem? (b) what can I do about it? (c) put a star on the best plan to pursue (d) do the plan, and (e) did the plan work? To evaluate this situation, one may use the seven (7) QC management tools, which allow the team to study the relationships. The tools are shown in Table 7.4.

PROCESS IMPROVEMENT CYCLE

There is a huge difference between *problem-solving methodology* and *process improvement.* A problem-solving methodology is a systematic approach to solving a particular problem. It assumes that there is an issue, concern, or problem that the customer (internal or external) is not pleased with, and it is the responsibility of whoever has that issue, concern, or problem to fix it. On the other hand, process improvement deals with the insatiable desire to make the process better than what it is now. It is imperative, however, that the process improvement is also systematic and structured. The reason for this is that it:

- Provides a more tightly focused process than the PSC, concentrating on improving the quality of a single process output (product or service).
- Lends itself especially well to customer issues, since it builds and strengthens ties between customer and supplier.
- Provides a catalyst for never-ending improvement through its continuous refocusing on customer needs and expectations.
- Affords a prime opportunity to utilize statistical process control (SPC) methods for process analysis, control, and improvement.
- Focuses improvement efforts on both product quality and waste reduction.
- Provides to improvement teams an enhanced motivational environment for their improvement efforts.

So, what is this structured process for improvement? Fundamentally, it is a 14-step process. A clarification and requirements of each step are as follows:

Step 1. Identify the output: Prepare a statement which answers the question, "What is to be done?" and consists of two parts: A tangible, specific noun followed by a verb describing what you do to produce the output. The statement should be neither too broad nor too specific, if it is to be useful. Examples: Shaft machined, doughnut fried, muffler installed, bicycle assembled, forecast developed, order taken, machine repaired, order shipped, report typed, *etc.*

Step 2. Identify the customer: The (primary) customer is the next person in the work process, the person who will receive your output as input and act on it. Therefore, "the" customer may be either external or internal. Secondary customers and end-users may be identified if their requirements are of significance.

Step 3. Identify the customer requirements: What does the customer want, need, or expect of the output? Some considerations are:

- Customer requirements may be general or very specific.
- Customer requirements often fall into such categories as: Cost, accuracy, timeliness, quantity, completeness, dimension, performance, appearance, *etc.*
- Supplier-customer interaction is essential. Both customer and supplier specify, negotiate, and evaluate information before reaching an agreement. Examples: Report needed ASAP; Expense cannot exceed the budget; Heat shield must withstand operating temperatures; Colors must match; Quick response time; *etc.*

Step 4. Translate requirements into product specifications: Applicable actions may be to:

- Customer's needs are put into language that reflects the supplier's active participation. [A good application tool is quality function deployment (QFD).]
- Specifications should be measurable, realistic, and achievable, and should be reviewed with the customer to be certain both parties understand and agree on what the output should be. Examples: Need by 4:00 pm Friday; OD 1.625 +/– .005"; Weight 5,200 pounds maximum; Typing double-spaced; *etc.*

Step 5. Identify the work process: What are the process elements? Applicable actions may be to:

- Use a fishbone diagram as a process analysis map (PAM) may help to identify the machines (equipment), measurement, manpower (people), material (input), methods, and environmental factors in the work process.
- List step-by-step what must be done to achieve the output. Flow charts are especially helpful here.
- Identify or develop the needed documentation that describes or provides instructions for the work process.

Step 6. Select the process measurements: Applicable actions may be to:

- Select the critical measurements for the process output. These measurements should be derived from customer requirements and product specifications and should permit an objective evaluation of the quality of output.
- Measurements should provide for early identification of potential as well as actual problems, where emphasis (stress) is on the prevention rather than on the detection of errors or defects. Examples: Cylinder OD in inches; Cake weight in grams; Transit time in days; Response time in seconds; Number of cycles before failure; Circuit resistance in ohms; *etc.*
- Also, select measurements for key process inputs (KPI), such as raw materials, incoming parts, energy, supplies, information, *etc.* In-process factors (intermediate process outputs), *i.e.*, in-process conditions that impact the process output, such as temperature, pressure, time, *etc.* Process performance criteria, such as scrap, rework, downtime, yield, delays, warranty, efficiency, *etc.*
- These measurements become the basis for process analysis, control, and improvement.

Step 7. Determine the process capability: This step helps determine if the existing process can produce output that consistently meets the agreed-upon product specifications with minimal waste. The process is operated according to standard practices; output measurement data are collected and analyzed to provide a statistical evaluation of process control (consistency) and process capability (ability to meet specifications). Applicable action may be to:

- Use statistical process control (SPC).

Step 8. Decide: Is the process capable? If the evaluation indicates that specifications are not being met consistently and that excessive waste (scrap, rework, delays, errors, *etc.*) is experienced, the work process must be revamped or improved. This requires problem solving; proceed to step 9. If the process capability assessment of step 7 determines that the existing process can produce output that consistently meets specification requirements and does so with minimal waste, proceed to step 10 to produce the output.

Step 9. Problem solved to improve the work process: Applicable actions may be to:

- Use the 5-step PSC to revamp or improve the existing process.
- If the process capability problem indicated in step 7 is a lack of process control (instability or inconsistency), indicated by sporadic deviations of the process average from the target, status quo, or historic level, the solution is one of identifying the destabilizing (special/assignable) causes and eliminating them. After the special cause has been identified, corrected, and is completed, loop back to steps 7 and 8 (reassess process control). If the problem is solved, proceed to step 10; if not, continue to repeat the PSC at step 9 as needed until process control is achieved.
- If the process capability problem indicated in step 7 is an inherent inability to meet specification requirements, the process system must be fundamentally changed. This requires identifying those key process factors (common or random causes) that have a major impact on the quality of process output and modifying them and their effects. After the common cause has been identified, corrected, and is completed, loop back to step 5 to redefine the revised work process and proceed. If at steps 7 and 8 the process capability problem is solved, proceed to step 10; if not, continue to repeat the PSC at step 9 as needed, until process capability is achieved. [A very good methodology to use for special cause is the DMAIC model of the six-sigma methodology.]
- If repeated attempts still leave you with an incapable process, talk with the customer who needs to know what to expect. A renegotiation of the requirements may be possible. If not, the customer may want to find another supplier.

Step 10. Produce the output: Produce the output, continuing to follow the standard practices (or the newly established practices which should be standardized for the new, revamped process).

Step 11. Evaluate the results. The focus here is "the" results and never on the process. So, Applicable actions may be to:

- Evaluation of results must be based on the product specifications that customer and supplier agreed to as part of step 4. Those specifications are a template against which results are compared.
- There is a major difference between this step and step 7 (determine process capability). Here you are evaluating how well you actually did (not how well you are capable of doing). The emphasis is on results rather than on process.

- There are two potential types of problems at this point: (1) the product output does not meet specification requirements, and (2) the product output meets specification requirements, but the customer is dissatisfied with the product.

Step 12. Decide: Is there a problem? Applicable actions may be to:

- If the evaluation of step 11 indicates that there is a problem of either type I or type II, problem solving must be carried out. Proceed to step 13.
- If no problem exists, continue production and proceed to step 14.

Step 13. Problem solved to improve results. Applicable actions may be to:

- Use the 5-step problem solving cycle (PSC) to troubleshoot the problem.
- If the problem with the process output is of type I, *i.e.*, the process output does not meet specification requirements, the cause of the problem is one of nonconformance to standard practices (since the process was shown to be capable in step). The solution to this type of problem lies in identifying the nonconforming practice (s) and restoring operations according to established standard practices at step 10.
- If the problem with the process output is of type II, *i.e.*, the product output meets specification requirements, but the customer is dissatisfied with the product, the cause of the problem is usually one of (a) improper understanding of customer requirements, (b) improper translation of requirements into specifications, or (c) improper process monitoring (measurement). Any of these causes requires looping back to step 3, 4, or 6 for corrective action.

Step 14. Recycle: At this point, there should be no evident problems of process capability or results. However, there may be opportunities for further quality improvement of the process output, for example, in terms of reduced variability. This could result in not merely meeting, but in exceeding customer expectations. And, of course, customer needs and expectations are rarely static. They are likely to change, especially in terms of becoming more restrictive. Applicable actions may be to:

- These opportunities for continuous process improvement require recycling back into the PIC.
- Under any circumstances, continue to monitor the work process and the results in order to maintain the required level of quality.

COMMENTS ON PROBLEM SOLVING VERSUS PROCESS IMPROVEMENT

One will notice that both the 5-step PSC and the 14-step PIC are proven, systematic approaches for continuous improvement. Both are extremely well-suited for team-oriented activity. However, there is always confusion as to what method to use. To be sure, it is not always obvious which of the two methods to use to tackle a particular work issue. It takes some skill and experience to make a better selection. The PSC/PIC selection chart, Table 7.5, compares the two methods in detail and provides an aid to judgment for making this decision. Fortunately, no matter which method you select, once you start using it, you are likely to learn quickly whether or not it is helping you to accomplish your objective.

TABLE 7.5
Problem Solving versus Process Improvement

Problem-Solving Cycle	Process-Improvement Cycle
A general method for making a change in: • systems • results • conditions • work process • management process	A tightly-focused method for ensuring conformance of: • a specific product • a specific service
This method fosters: • definition of problems • analysis of data • understanding of causes • creative ideas • more alternatives • teamwork • commitment	This method fosters: • elimination of unneeded work • prevention of problems • shared responsibility • strong customer/supplier communication lines • evaluation of work processes • confidence in results
Use this method when: • there is a gap between what is happening and what you want to happen • you want to move from a vague dissatisfaction to a solvable, clearly- defined problem • you're not sure how to approach an issue	Use this method when: • you need to improve the quality of a particular, currently-existing output • you don't have agreed-upon customer requirements for an output • you are about to produce a new output, the need for which has recently been determined
Switch to the PIC when: • the problem you identify is a lack of quality or an inability to assess quality • the recommended solution involves producing a specific output	Switch to the PSC when: • evaluation of Process Capability shows that the current work process cannot produce a conforming output • evaluation of results shows that the work process did not produce a quality output

In general, use the PIC when you need to improve the quality of a particular, currently existing output, or you are about to produce a new output. Use the PSC when there is a gap between what is currently happening and what you want to happen. Note that the PSC is an integral and vital part of the PIC (at steps 9 and 13).

We mentioned that both approaches are well-suited for team-oriented activities. The question then becomes: Why is it necessary to use teams? Because when we deal with problems, all of us are better than any one of us. In fact, if we look at the word TEAM as an acronym, we will find the answer. **TEAM** = **T**ogether **E**veryone **A**ccomplishes **M**ore. This is the essence of synergy and consensus. Specifically, management has recognized that the goal of any employee involvement (EI) is to give people increased influence over and responsibility for their own work. In recent years, one of the major routes to achieving this goal has been the creation of problem-solving teams: Groups of employees who meet on a regular basis to identify and solve work-related problems. They are indeed realizing that they have a vast, untapped potential in their employees' minds – especially for their own processes. To remain competitive, it is essential to make effective use of all the talents that our labor force has to offer, both physical and mental. Successful companies worldwide credit much of their success to the widespread use of employee teams to generate improvements in all departments and at all levels.

Quality improvement in any CS endeavor is hard work and takes a long time. It is all too easy for one person's commitment and enthusiasm to flag during a long project. The synergy that comes from people working together toward a common objective is usually enough to sustain the enthusiasm and support. Therefore:

- We need to provide employees with more challenging activities to keep their active minds fully utilized. Most employees want to participate in the decision-making and problem solving that affect them. All the people affected by a problem should be a part of the solution. This assures a "buy-in" to the corrective action or solution put into place.
- We must recognize that no one individual has all the process and product knowledge plus the special skills and experience required for optimal problem solving.
- The most effective, proven, problem-solving, and process-improvement methods and tools lend themselves best to a team approach.
- A team will invariably generate more problems, more causes, and more solutions than any individual can.

As important as the teams are, we must also recognize that there are two types of teams. The first is the local, and the second is the cross-functional team.

Local, or Department Improvement Teams (DITs) are comprised of all the members of a department. These employees typically work in close proximity, experience common problems, and form a "natural work group." Their purpose is to provide a focus and a means for all employees to contribute to an ongoing activity aimed at improving the quality and productivity levels of the department.

On the other hand, cross-functional, or Process Improvement Teams (PITs) are created to continuously improve quality, reduce waste, reduce warranty, and improve productivity of a process that crosses many departmental lines. The PIT is made up of experienced, skilled problem solvers from all departments involved in and affected by the process. A typical PIT is the six-sigma team that addresses major problems for the organization.

For either team to be effective, they require at least the following:

- A team charter, which specifies the team's purpose and enumerates the team's duties and responsibilities.
- Selection of the proper team makeup.
- Selection of an effective team leader and team coordinator.

- Knowledge of the organization's mission, goals, and objectives.
- Learning to work together as a team ("team building").
- Adequate training in problem solving and process improvement methods and tools.
- Guidelines and ground rules for holding effective meetings and for making decisions.
- An adequate meeting place with needed support services, such as availability of computers, typing, copying, information, *etc.*
- Use of meeting agendas and minutes (recaps).
- Liberal credit and recognition for team successes.

So, what does it take to solve a problem? We believe that two essential attributes must be present. The first is an appropriate problem-solving methodology that supports the organizational goals (preferably standardized to reflect organizational needs), and the second is an appropriate team (members must have appropriate skills, ownership, and knowledge) for the task. The team must be part of an organizational culture that encourages:

- A climate/environment of trust.
- Encourage an atmosphere of reasonable risk.
- A commitment to common goals.
- Honest and open communications.
- Common agreement on high expectations for the team.
- Assumed responsibility for work that must be done.
- A general feeling that one can influence what happens.
- Common access to information.
- A win-win approach to conflict management.
- Support for decisions that are made.
- A focus on process as well as results.

8 Eight Disciplines (8D)

OVERVIEW

In Chapter 7, we discussed the issue of problem-solving methodologies, and the 8D was mentioned as one of the common approaches (in many cases, the preferred way) to effectively resolve problems. In this chapter, we are going to discuss this particular methodology in more detail because it is used very frequently and by many suppliers in many different industries. Fundamentally, the methodology is actually a nine-step process, even though it is identified as an eight-step approach. It was developed by Ford Motor Company in the mid to late 1980s and has been revised several times since then (Ford Motor Company, 1998). The final revision added one more step, and that made it a nine-step approach. However, since everyone was accustomed to 8D, Ford decided to rename it global 8D (G8D) to reflect the new approach. The new approach has very few changes other than the addition of the D0 step. We have chosen to use the old designation of 8D only because that is the name that most of the quality professionals are accustomed to, even though in reality the G8D methodology is used. So, the goal of the 8D, as we will explain it, is equal to the G8D, and its purpose is to eliminate defects so that they do not recur. A defect, in the context of the 8D, is: A problem, a concern, an opportunity. The steps are identified as D0-D8, which stand for D = discipline, and the numerical numbers 0-8 are the actual steps. In a summary form, the steps are:

- **D0**: *Prepare for the 8D process*: In response to a symptom, evaluate the need for the 8D process. If necessary, provide Emergency Response Action (ERA) to protect the customer and initiate the 8D process.
- **D1**: *Establish team*: Establish a small group of people with the process and/or product knowledge, allocated time, authority, and skill in the required technical disciplines to solve the problem and implement corrective actions. The group must have a designated champion and team leader. The group begins the team-building process.
- **D2**: *Describe the problem*: Describe the internal/external customer problem by identifying "what is wrong with what" and detail the problem in quantifiable terms. The more precise the operational definition of the problem is, the higher the probability of solving the problem.
- **D3**: *Develop the interim containment action (ICA)*: Define, verify, and implement the ICA to isolate effects of the problem from any internal/external customer until permanent corrective actions (PCAs) are implemented. Validate the effectiveness of the containment actions.
- **D4**: *Diagnose problem*: Define and verify root cause and escape point. Isolate and verify the root cause by testing each possible cause against the problem description and test data. Also, isolate and verify the place in the process where the effect of the root cause should have been detected and contained (escape point), but was not.

DOI: 10.1201/9781003650614-8

- **D5**: *Choose and verify permanent corrective actions (PCAs) for root cause and escape point*: Select the best permanent corrective action to remove the root cause. Also, select the best PCA to contain the effect of the root cause. Verify that both decisions will be successful when implemented without causing undesirable effects.
- **D6**: *Implement and validate permanent corrective actions (PCAs)*: Plan and implement selected PCAs. Remove the ICA. Validate the actions and monitor the long-term results.
- **D7**: *Prevent recurrence*: Modify the necessary systems, including policies, practices, and procedures, to prevent recurrence of this problem and similar ones. Make recommendations for systemic improvements, as necessary.
- **D8**: *Recognize team and individual contributions*: Complete the team experience, sincerely recognize both team and individual contributions, and celebrate.

[**Special note**: The 8D methodology is an eight-step, team-based approach to solving problems, while DMAIC (Define, Measure, Analyze, Improve and Control), born from the Six Sigma methodology, is a five-step, data-driven process for improvement and project management. The two methodologies are very powerful and useful if applied for the intended purpose. On the other hand, the DMADV (Define, Measure, Analyze, Design and Verify) prevents defects (proactive). DMAIC includes specific solutions; DMADV is part of the solution design process.]

8D APPLICATION CRITERIA

Contrary to many practitioners 8D is not necessary in all situations. There are six definite guidelines when the 8D methodology is appropriate and applicable. They are:

1. There is a definition of the symptom(s). (The symptom has been quantified.)
2. The 8D customer(s) who experienced the symptom(s), and when appropriate, the affected parties have been identified.
3. Measurements taken to quantify the symptom(s) demonstrate that a performance gap exists AND OR priority (severity, urgency, growth) of the symptom warrants initiation of the process.
4. The cause is unknown.
5. Management is committed to dedicating the necessary resources to fix the problem at the root cause level and to prevent recurrence.
6. Symptom complexity exceeds the ability of one person to resolve.

If any one of these items is present, then an 8D is appropriate. All 8Ds require some common tasks that optimize the resolution of the problem. These common tasks need continual consideration throughout the problem-solving process using the 8D approach. Some of these considerations have to do with making sure that the team is responsible to:

- Document the changes.
- Review team composition.

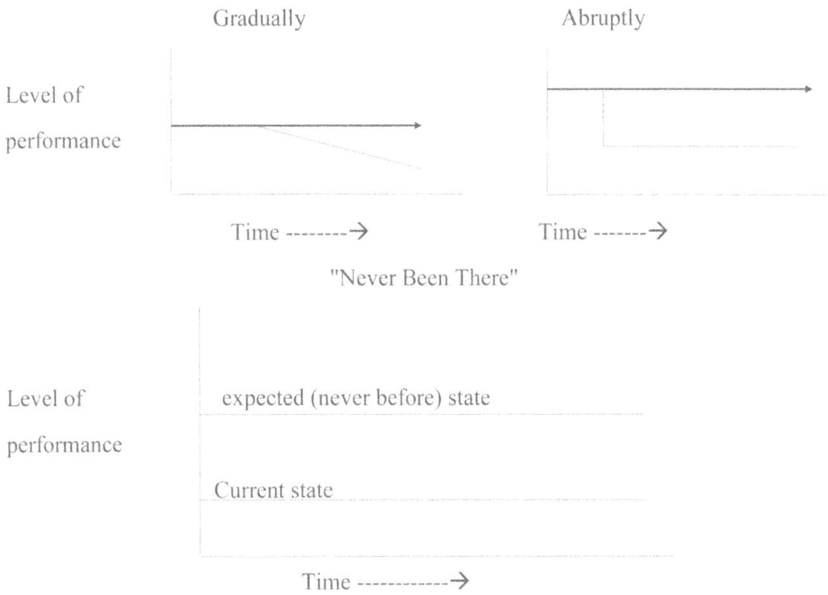

FIGURE 8.1 A pictorial view of the three possibilities of something changed.

- Review measurable(s).
- Determine if service action is required.
- Review assessing questions.
- Update the 8D report itself.

Recognize that a change has occurred. There are three possibilities: (a) Something has changed gradually. (b) Something has changed abruptly, and (c) Something new is expected. Never been there before. These changes are shown in Figure 8.1. Now that we know that a change has occurred, let us address each of the steps with a little more detail.

D0: Prepare for the Problem-Solving Process

In response to a symptom, evaluate the need for the 8D process. If necessary, provide ERA to protect the customer and initiate the 8D process.

The application criteria identifies conditions that justify the use of the 8D process. This justification is necessary to allocate resources because the demand for resources in terms of people, time, money, experience, *etc.*, is usually greater than what is available.

Many problem-solving techniques and processes are available. Matching the right problem with the right process is one way to avoid wasting limited resources. Using those resources effectively is also important. The 8D requires the use of many resources, but it offers many benefits. Of special interest here is that it is the choice of the supplier to use whatever method they want to use for identifying and correcting the defect. However, some customers demand (require) that the results of the supplier's investigation must be in the customer's 8D format.

In this stage, key questions that must be addressed are of two kinds. The first one deals with emergency response (containment) action. Typical questions may be:

- Are ERA necessary?
- Is a service action required as part of the emergency response?
- How was the ERA verified?
- How was the ERA validated?

The second kind deals with the application criteria. Typical questions here are:

- Will the new 8D duplicate an existing 8D?
- How well does the proposed 8D meet the application criteria?
- Is there a definition of the symptom(s)? Has the symptom been quantified?
- Has the 8D customer(s) who experienced the symptom(s) and, when appropriate, the affected parties, been identified?
- Have measurements been taken to quantify the symptom(s) demonstrated that a performance gap exists AND/OR has the priority (severity, urgency, growth) of the symptom warranted initiation of the process?
- Is the cause unknown?
- Is management committed to dedicating the necessary resources to fix the problem at the root cause level and to prevent recurrence?
- Does the symptom complexity exceed the ability of one person to resolve the problem?

In some cases, additional basic questions may have to be asked, such as:

- Have all changes been documented (*e.g.*, FMEA, control plan, process flow)?
- Do we have the right team composition to proceed to the next step?
- Have we reviewed the measurable(s)?
- Have we determined if a service action is required?

D1: Establish the Team – Process Flow

The purpose of this step is to establish a small group of people with the process and/or product knowledge, allocated time, authority, and skill in the required technical disciplines to solve the problem and implement corrective actions. The group must have a designated champion and team leader. The group begins the team-building process.

It is imperative that everyone understands that the original "group" must develop into a "team" for an effective problem resolution. This means that the team must go through the traditional development of forming, storming, norming, performing, and adjourning (Tuckman, 1965; Blanchard and Parisi-Moger, 2009; Abudi, 2010).

The reason for team development is that problem-solving implies changes will be made to resolve an *undesired condition*, which implies performance based on synergy. Comprehensive or synergistic problem-solving approaches acknowledge the need to make provisions for the entire problem-solving effort. Such provisions may include blending the organization's needs with the problem solvers' skills, behaviors, and personality styles.

Step D1 addresses part of the requirements for successful problem-solving efforts. Of course, some problems can be tackled by one person, but generally speaking, resolving a concern at the systemic level requires a cross-functional group, a team.

A problem-solving team may be created with specific tasks or objectives in mind. However, when people with different experiences, skills, knowledge, and priorities are assembled, human relations issues tend to arise and interfere with the completion of assigned tasks.

D1's primary function is to prevent these human relations issues from developing. During D1, roles, goals, and responsibilities are clarified. In group problem-solving situations, the team is well advised to establish guidelines for group and individual behaviors. At a minimum, the owner of the concern (called the champion), the one with decision-making authority, must be strongly committed to the problem-solving effort. Fixing the problem at the root cause level and subsequent prevention efforts are impossible without this commitment. The champion's critical role is to empower the problem solver(s) to complete the prescribed problem-solving steps. The champion also helps remove barriers and procures resources needed to resolve the concern.

Team Composition

For the team to function properly and effectively, there are specific roles and responsibilities that the participants must have. Some of the key ones are:

Champion: The champion is the person who has the authority to make (and implement) decisions regarding containment actions, PCAs, and prevention of recurrence. He or she also supports the problem-solving efforts of the team throughout the process. The champion also sets priorities of specific tasks, as well as being the individual with power and influence to diffuse any bottlenecks that may occur during the problem-solving process. Some of the key duties are:

- Tasks the team with the 8D method.
- Helps remove organizational barriers to the 8D method.
- Helps procure resources the team requires to complete the process.
- Reviews assessing questions with the team.
- Exercises authority to implement team recommendations.
- Supports the organization as a point of contact regarding the 8D process.
- Forwards the 8D report to concerned departments.

Team Members: Team members are those who are selected to participate on the 8D team due to their expertise and knowledge. They are the subject matter experts who do the work of each objective. The members might change during the 8D process. Generally speaking, they are responsible for:

- Doing the investigative work.
- Developing plans.
- Using their judgment, skills, experience, and knowledge.
- Finding the answers.
- Developing recommendations at each 8D step.
- Executing the implementation actions.

Team Leader: The team leader is the individual responsible for leading team members through the 8D process. This person has leadership and interpersonal skills. The team leader is not the chairperson of the team. This is a person who makes sure that synergy is cultivated for the optimum solution. Typically, the team leader:

- Leads the teams to complete each 8D objective.
- Develops agendas for meetings and team activities.
- Conducts or leads the meetings.
- Asks (not generally answers) the 8D process questions.
- Manages the team in accordance with established team effectiveness guidelines.
- Supports the champion as the primary point of contact.

Recorder: The recorder is a team member who generates, holds, and publishes team reports (such as 8D reports, meeting minutes, agendas, action plans, *etc.*). Typically, the recorder:

- Controls documents for the team.
- Tends to be an administrative support person for the team.
- Takes responsibility for creating and distributing meeting minutes and reports in a timely manner.

Facilitator: The facilitator's primary task is to assist the team with interpersonal relationship issues such as resolving conflict, arriving at consensus (when appropriate), confronting counterproductive team behaviors, *etc.* The facilitator's role is optional and depends on the style of the leader. If the leader is not open-minded to open discussion, the leader is biased toward a particular approach, the leader does not allow for 100% participation by everyone, then a facilitator is necessary. Typically, the facilitator:

- Works with the team to resolve conflicts.
- Shows team members ways to work together while maintaining individual self-esteem.
- Coaches individual team members, leaders, and champions on how to be more effective in team situations.
- Helps the team develop team effectiveness guidelines.
- Provides feedback to the team on team performance behaviors.
- Helps the team apply techniques such as the decision-making process, creative processes, *etc.*, but is neutral in decision-making situations.

D1: Process Guidelines

The purpose of D1 is to form and develop the team with the required experience for the 8D process. During D1, goals, roles, and responsibilities are developed and defined. However, the type of problem usually defines what experience and skills are needed and, therefore, who needs to be on the team.

- All members must contribute, use their experience, and help gather facts.
- Team members may change, as the team's required experience changes. An ideal team size is about seven, and generally, but not necessarily, ranges from five to nine members to avoid ties.

- Team positions include leader, recorder, members, and facilitator.
- The roles on an 8D team are interdependent. No one can do it by himself/herself.
- The champion is the person who owns the problem, supports the team's efforts, uses authority and influence to help the team with organizational barriers, and helps get resources that the team needs to complete each 8D step.
- The entire 8D team works with the champion's help, involvement, and agreement.
- The champion should conduct the first meeting to describe what is needed and why the problem needs to be resolved, and to outline what is expected of the team.
- The leader is expected to lead the team, not have all the answers. Critical competencies for the leader are people handling and leadership skills.

At D1, there are more questions than definitive answers about the problem. This is normal. Better data regarding the problem will be created as the team progresses through the 8D process. At D1, the team members, champion, and others are selected using the best currently available information. For an optimum selection of team members, there are several key questions that should be addressed before the closure of this step. They deal with logistical issues (warming up) as well as membership questions. Typical ones are:

Warming Up

- What have you done to make the room user-friendly?
- When and where will the team meet?
- What has been done to help team members build their relationships with each other?
- What has been done to help team members focus on the team's activity?
- Has the purpose of the meeting been stated?
- Has the team been informed of the agenda for the meeting?

Membership

- Are the people affected by the problem represented?
- How is the 8D customer's viewpoint represented?
- Does each person have a reason for being on the team?
- Is the team large enough to include all necessary input, but small enough to act effectively?
- Does the team membership reflect the problem's current status?
- Do the team members agree on membership?
- What special skills or experience will the team require in order to function effectively?
- Have the team's goals and membership roles been clarified?
- Does the team have sufficient decision-making authority to accomplish its goals?

- How will the team's information be communicated internally and externally?
- Do all members agree with and understand the team's goals?
- Has the designated champion of the team been identified?
- Has the team leader been identified?
- Are team members' roles and responsibilities clear?
- Is a facilitator needed to coach the process and manage team consensus?

D2: Describe the Problem

D2's purpose is to describe the internal/external customer problem by identifying "what is wrong with what" and detailing the problem in quantifiable terms. The word problem may refer to the cause, concern, defect, or consequences of the defect's occurrence. Therefore, the word problem is too general a term. That is why D2 makes a clear distinction between the cause and the effect. The function of D2 is to factually describe the problem with a correct operational definition. In 8D terminology, the problem is a deviation from expectation, a special cause, a distribution within control limits that are too wide, or any unwanted effect where the cause is not known.

In order to factually describe the problem, we need data. The 8D process is a data-driven process. The focus of D2 is to obtain an accurate and unbiased description of the object and defect. The information collected during D2 is critical to successful problem resolution. Historically, problem-solving proponents have advocated that defects be described using either rationally based or experientially based techniques. Each approach has advantages and limitations. The 8D process advocates a synergistic approach to problem description. The result is a factually based, team-generated problem description.

To generate a fact-based approach, it is necessary to allocate the appropriate time for understanding and gathering the appropriate and applicable data. The time spent in D2 is always recovered (manyfold) by the time saved in subsequent 8D steps. Unfortunately, this feature of D2 is not obvious until later in the process. Teams that quickly pass through or skip D2 (and so have not adequately described the problem) have solved the wrong problem.

The initial D2 step develops a problem statement, which is then expanded by the factual listing of the problem's profile. Critical features of the problem are made visible using an accurate process flow diagram, which depicts the entire process (including any existing informal fixes). This body of information is further enhanced by developing a complete cause-and-effect (fishbone) diagram. Supplemental, effect-specific techniques (often experience-based) are useful.

D2 includes various types of reports and description approaches. Completing all these fact-gathering techniques may seem excessive. In reality, the synergism created by combining these techniques is of enormous benefit from D3 through D7. These remaining objectives tend to be realized more quickly, more effectively, and with far less frustration.

A significant byproduct of this synergism is the positive effect on the team members' behaviors. Team members appreciate each member's experience, resist the tendency to jump to conclusions, and realize the benefits of teamwork. The need for a

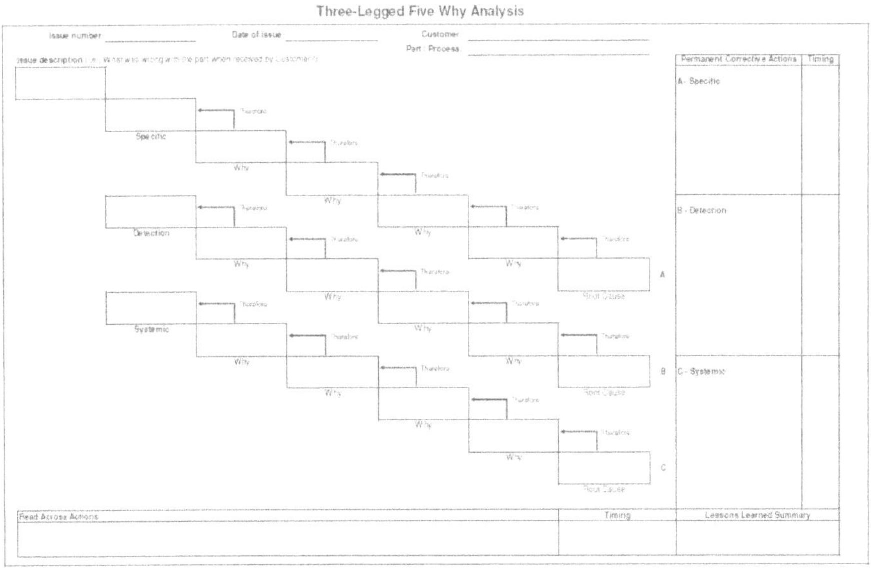

FIGURE 8.2 Three legged five why analysis format.

leader remains, but the team is prepared to proceed from a complete and factual (not opinionated) foundation.

In this step, the team may want to use the 3 × 5 why method – Figure 8.2. In Chapter 7, we just mentioned it and gave a pictorial view of it. Here we show the reader a simpler form with some of the characteristics. The 3 Legged 5 Why is a much more focused approach to problem solving, if we focus on the underlying root cause (as specific as possible), systemic issues or the escape (detection mechanism). Specifically it focuses on key generic questions, such as:

- How was it created?
- How did we miss it?
- How did our system fail?
- Ask, "Why?" as many times as necessary (usually FIVE times).
- Check your analysis with the "therefore" test – in a backward flow.

Depending on the answers, the analysis goes further, dealing with the following:

- Direct root cause: How did the problem occur?
  - Start with the actual problem.
  - Keep the focus simple.
- Detection root cause: How did the problem escape?
  - Focus on inspection system.
  - How did it fail?
  - Did we account for problem in our Detection Processes?

- Systemic root cause: Why didn't our systems protect the customer?
    - Start with the Specific Root Cause
    - Look at the problem from a bigger perspective.
    - Find organizational flaws.

Obviously, these questions are only a guideline. In the real process, there will be more in both volume and depth. Here we are emphasizing that this specific method allows the team to: (a) drives corrective actions for all 3 legs (b) separates "noise" from real issues (c) forces people to "dig deep" into processes, procedures, data (d) makes management issues visible, and (e) provides focus for the efforts of the cross-functional team action, and systemic issues.

D2: Process Guidelines

The purpose of D2 is to define the problem for which you eventually want to determine the root cause. D2 is not intended to determine the root cause. The information will be used later (at D4) to determine the root cause. Specifically,

- D2 is the observation collection step.
- It establishes the data that is the foundation for all other 8D steps.
- It demands the team develop:
    - A problem statement.
    - A problem description using IS and IS NOTs (what, where, when, how big?). See Table 8.1.
    - A process flow diagram.
    - A cause-and-effect diagram (also known as a fishbone diagram).
    - The problem may also need to be defined using additional approaches (*e.g.*, SPC data, customer satisfaction, and field reports).
    - To complete this step, the team may need to obtain missing information.
    - Recommended attachments to the 8D Report at D2 include IS/IS NOTs, 3 × 5 Why and the cause-and-effect diagram.
    - The team and champion use their judgment to determine how much information is sufficient.
    - The process flow diagram should include the initial actions taken to compensate for the problem when it first occurred.
    - Note all temporary or informal fixes on your process flow. Do not remove them, just note them.
    - Start where the defect can be first seen, then work your way back from there to the beginning of the process.

Use only one 8D per problem. Do not use a single 8D on many different problems. Each problem probably has a different root cause.

How to Develop a Problem Description

To develop the correct problem statement, there are at least two fundamental issues that have to be answered. The first is: "What's wrong with what?" If the answer is "No" the answer to "What's wrong with what" is your problem statement. If the answer is "Yes" repeat the question, "What's wrong with what" until reaching the statement where the cause is unknown. To develop the problem description, apply

TABLE 8.1
A Typical Format for the IS/IS NOT

IS	IS NOT
What	**What**
Name the object that is having trouble. Name the trouble or problem that the object is having	Name objects that are similar in shape, composition, form, or function that could have the same trouble but are not. Name other kinds of troubles or problems this object could be experiencing, but is not experiencing.
Where	**Where**
Name the place where this object with the trouble can be found. Name the place where this problem first showed up. Name all other places where this problem occurs (or has occurred). Where on the object is the problem occurring (inside, outside, top, bottom, *etc.*) Where in the process flow did the problem first develop Describe all other places with regard to where the object and the trouble can be found	Name other places where this object can be found without the trouble. Name places where the problem could first have showed up but did not. Name similar places where this problem has never occurred. Where on the object could the problem occur, but is not. Where in the process could the problem have first developed, but did not? Use similar units of measure to describe where the object and the trouble are absent.
When	**When**
When with respect to time did the trouble/problem first occur Describe the above information with regard to day, month, year, time of day, *etc.* Describe any patterns with regard to time Describe when in the process the trouble/problem first occurred. Describe when in the life cycle the problem first occurred. Describe other places in the process and life cycle where the trouble or problem was observed. Note: Consider all units of time, such as hours, days, minutes, seconds, shifts, periods, quarters, years, *etc*. Also consider sequences in operations as would occur in a process flow diagram	Using the same unit of time, when could the trouble and or problem have first occurred, but did not. Describe the above information with regards to matching day, month, year, time of day, *etc.* when the trouble did not occur. Describe corresponding units of measure to describe when the trouble did not occur. Describe when the trouble/problem could have first occurred, but did not. Describe when in the life cycle the problem could have first occurred, but did not. Describe when it was not.
How Big	**How Big**
Describe the size of the problem/trouble/effect Describe the number of objects that have or have had the trouble/problem/effect.	Using a similar unit of measure describe the limits of the problem/trouble/effect. Describe the number that could have had the trouble/problem/effect, but did not.

(Continued)

TABLE 8.1 (*Continued*)
A Typical Format for the IS/IS NOT

IS	IS NOT
Describe the magnitude of the trouble in terms of percentages, rates, patterns, trends, yield, dimensions, *etc*.	Describe what it could have been, but is not.
Describe the number of defects per object.	Describe what it could have been, but is not.
Describe the physical dimensions of the defect or problem.	Describe in physical dimensions the defect that could be demonstrated, but has not.
ASK:	
"Are there any other facts that do not appear on the is/is not tool?"	
"Who can we talk to about the is/ is not facts?"	

the IS/IS NOT method using the format in Table 8.1. The second is: "Do we know why this is happening?" Once you know the answers to these two questions now the process for developing the problem description is as follows.

Step 1. Develop the problem statement: Develop a simple, concise statement that describes who or what object shows the specific defect you want to determine the root cause of. Intended as a starting point for further description, the statement must be very concise and vivid. Use the repeated whys to produce the lowest rung in what might be a series of events, a chain reaction. This stage is very critical. As such, make sure that the symptom is not confused with the problem. Generally speaking, the customer identifies the *symptom*. It is up to the team (*i.e.*, engineering, quality, or even purchasing) to define the "root cause" of the problem.

Step 2. Develop the problem description: Describe in quantifiable, factual terms the extent of the defect. List all facts with regard to who, what, where, when, and how big (magnitude). Describe in quantifiable, contrasting dimensions where the symptom is not manifested or visible.

Step 3. Identify information which needs to be confirmed or collected: Obtain the information from knowledgeable people by asking the same questions used in step 2. Sometimes in conducting the IS/IS Not analysis, it is necessary to do a comparison analysis to identify differences in.

- Facts
- Unique items to the IS
- Facts that have not already been stated in the IS column.

To uncover any differences, the key question is: What is different, unique, peculiar, special, or true only of the IS when compared to the companion IS NOT? A good guideline to develop differences is to:

- List all facts without prejudice to cause or potential cause.
- Consider the categories of people, methods, materials, machines, measurement, and mother nature (environment).

- A second analysis that is often necessary is the comparison of changes in:
  - Facts
  - Specific items related to differences.

To uncover the changes, the key question is: What has changed in, on, around, or about the difference, and when did the change occur? A good guideline to develop changes is to:

- List all changes that occurred related to difference regardless of date or their potential for cause.
- Consider the categories of people, methods, materials, machines, measurement, and Mother Nature (environment).

Once the completion of the IS/IS Not analysis is finished, the team is ready to begin the identification and or the development of possible cause (Change-How Theories). At this stage, theories are:

- Statements of ways that the changes may have created the trouble.
- A limited form of brainstorming.
- Simply a listing of possible causes, not probable causes.

To develop theories, the key question is: How could this change have caused the effect on the object? (Or, in what ways might this change have caused the effect on the object?) A good guideline to develop change-how theories is to:

- Just list each theory individually.
- Do not reject or qualify the theory based on its practicality or probability.
- List single change/single variability theories, first.
- Develop more complex variability theories, second.
- Continue to prompt the problem-solving team with the above question until all possible theories are developed.
- Defer critical thinking until the next step.

At this stage, the team is ready to check the theories. This is done with trial runs, such as:

- Critical evaluations of theories against the sets of IS/IS NOT data.
- Test of the plausibility, not a remote possibility of each theory.
- A test of the likelihood of the theory.
- A process of elimination.

To test the theory, the key question is: Does the change, how theory explain both the IS and the IS NOT? A good guideline to develop change-how theories is to:

- Test the theory against each individual set of IS and IS NOT.
- If the theory accounts for both the IS and the IS NOT phenomena, record with a + (plus) symbol.

- If the theory cannot plausibly explain either the IS or the IS NOT phenomenon, record the result with a - (minus) symbol and note the reason for the rejection.
- Test the theory against all sets of IS and IS NOTs. Do not stop testing a theory prematurely.
- Test all the theories in and of themselves. Consider each "change-how" theory a separate theory.
- Avoid testing interactive changes and highly complex theories to the end of the trial runs. Test the simplest theories first.
- If the plausibility of the theory is uncertain (+ or –), use a question mark (?) to record the uncertainty. Note why the theory is uncertain.

This simple approach does not suggest that the team may not use any advanced statistical techniques. The team is free to use any methodology or any tool that may help solve the problem.

To finish the requirements of the D2, the last item should be to verify the root cause by:

- Describing the optimum method to passively, and then actively, verify the root cause.
- Conducting the verification in the appropriate setting.
- Recording the results.

D2: Evaluating Questions

When the team feels comfortable of having completed the D2 step, it is a good practice to review the discussion with at least the following basic questions:

- Can the symptom be subdivided?
- Has a specific problem statement been defined (object and defect)?
- Have repeated whys been used?
- What's wrong with what?
- Do we know for certain why that is occurring?
- Has Is/Is-Not analysis been performed (what, where, when, how big)?
- When has this problem appeared before?
- Where in this process does this problem first appear?
- What, if any, pattern(s) is (are) there to this problem?
- Are similar components and/or parts showing the same problem?
- Has the current process flow been identified? Does this process flow represent a change?
- Have all required data been collected and analyzed?
- How does the ERA affect the data?
- Is there enough information to evaluate for potential root cause?
- Do we have physical evidence of the problem?
- Has a cause -and-effect diagram been completed?
- Does this problem describe a "something changed" or a "never been there" situation?

- Has the problem description been reviewed for completeness with the 8D customer and affected parties? (Has concurrence been obtained from the 8D customer and the affected parties?)
- Should this problem be reviewed with executive management?
- Should financial reserves be set aside?
- Should any moral, social, or legal obligations related to this problem be considered?

D3: Develop the Interim Containment Action (ICA)

The D3's purpose is to define, verify, and implement the ICA to isolate the effects of the problem from any internal/external customer until PCAs in D6 are implemented. Validate the effectiveness of the containment actions. In other words, the D3's predominant function is to protect the customer from the problem until it is resolved at D6.

The customer is whoever gets your output as their input. The customer could be internal to the company as well as external. If the customer cannot accept the defect(s), then an ICA is required. An ICA is intended to protect the customer from getting the problem because the root cause is not yet known. It must be understood that an ICA is always a temporary measure. It is an added non-value operation. Commitment to completing the 8D remains in force. An ICA is effective when, from the customer's point of view, the defect is no longer evident. ICAs are often compared to quick fixes or hidden factories, but containment actions are actually:

- Verified to work before implementation.
- Validated that the ICA is working after implementation.
- Validated independently, in addition to the customer's confirmation of effectiveness.
- Documented to exist (*i.e.*, made visible) by amendments to the organization's process flow diagrams, procedures, process instructions, *etc.*
- Replaced by a permanent corrective action (PCA) at D6.
- Implemented without creating other troubles downstream.
- Formal temporary fixes. Care is exercised to ensure that all supporting functional areas/departments are involved in planning and implementation.
- Implemented only with the clear approval of the champion and customer.
- Managed through comprehensive planning and follow-up.

Another important feature of D3 is that prudence may require containment action (an ERA) before Dl and D2 are completed. However, after D2 is completed, the ERA should be reviewed. The D2 data may suggest a more effective ICA.

An ICA should not be implemented by a team without the champion's knowledge or involvement. A subtle but significant feature of the 8D process is that the authority to implement an ICA is retained by the champion. It is not delegated to the team. Likewise, the champion is responsible for cross-functional communications. Some team problem-solving guidelines imply that the team itself would assume this responsibility. The 8D process promotes action within the organization's chain of command. It does not promote creation of an artificial command structure which bypasses the normal authority structure.

D3: Process Guidelines

As mentioned earlier, the fundamental purpose of D3 is to protect the customer from receiving a defective part(s) while the supplier is determining the source of the cause. Steps and questions during D3 include:

- Is D3 step necessary? Sometimes it may not especially if the ERA is sufficient to control the product.
- If implemented, can the D0 ERA be improved?
- Which action(s) is/are best?
- Choose the best containment action.
- Will the containment action work (*i.e.*, eliminate 100% of the problem from reaching the customer)? (Is verification valid and accurate?).
- Follow the management cycle – PDCA cycle.
- Plan (in detail) the containment action(s) and verify.
- Implement the containment action.
- Record the results.
- Evaluate the ICA; did it work? (Validation).
- Document the actions and communicate how to do it to all who need to know about it.
- Continue to monitor its effectiveness throughout the time it is in place.
- The champion must be involved before the ICA is implemented and during implementation.
- Is the only optional step in the entire 8D process.
- Is temporary, not permanent. It is costly, not cost-effective. It is containment, not mistake-proof.
- Works against the problem, not the root cause.
- Increases costs while the ICA is used.
- May be done while other members of the team are working on D4 (determine root cause).
- May require additional members to be added to the team in order to complete all of D3.
- Looks like a quick fix, but its actions are documented, later removed, and checked for effectiveness. The team must ensure that the actions will not create other problems for departments and customers downstream of the D3 action(s).
- More than one D3 action may be required to fully protect the customer (internal or external).

After D3, the team's membership might change. This is true at all steps. May occur before Dl or D2. If done before D2, then the D3 ICA should be reviewed after the D2 data is compiled. Better D2 data may prescribe the need for an improved D3 ICA.

D3: Evaluating Questions

When the team feels comfortable of having completed the D3 step, it is a good practice to review the discussion with at least the following basic questions:

- Are ICAs required?
- Is a service action required as part of the ICA?

- What can we learn from the ERA that will help in the selection of the "best" ICA?
- Based on consultation with the 8D customer and champion, have criteria been established for ICA selection?
- Based on the criteria established, does the ICA provide the best balance of benefits and risks?
- How does this choice satisfy the following conditions?
    - The ICA protects the customer 100% from the effect.
    - The ICA is verified.
    - The ICA is cost-effective and easy to implement.
- Have the appropriate departments been involved in the planning of this decision?
- Have appropriate Advanced Product Quality Planning (APQP) tools (*e.g.*, FMEA, control plans, instructions) been considered?
- Have plans, including action steps, been identified (who needs to do what by when)?
- Has a validation method been determined?
- Does the customer have a concern with this ICA (is customer approval required)?
- Have we identified what could go wrong with our plan, and have preventive and contingency actions been considered?
- Are implementation resources adequate? This is a very important consideration, especially when third-party auditors (inspection houses) do the ICA.
- Does the validation data indicate that the 8D customer is being protected?
- Can the ICA effectiveness be improved?

D4: Define and Verify Root Cause and Escape Point

D4's purpose is twofold. The first is to isolate and verify the root cause by testing each possible cause against the problem description and test data. The second is to isolate and verify the place in the process where the effect of the root cause should have been detected and contained (escape point), but was not.

To appreciate the rationale of the D4, we must understand that a situation was once problem-free, but now is not. When this happens, then the situation is considered a change-induced condition. The process of identifying this change is called problem solving. Various problem-solving techniques exist – Appendix B. However, they are all based on three basic strategies:

- Fact-based, deductive approaches.
- Experientially based approaches.
- Creative approaches.

A working knowledge of various problem-solving techniques is desirable. The single verified reason that accounts for the problem is the root cause. It is the cause that explains all facts about the problem. It is verified by the ability to make the problem come and go on demand.

Determining Root Cause

Information compiled at D2 is used to identify a set of possible causes. At D4, the root cause is discovered. Problem-solving techniques are used to reduce the time and confusion to systematically deduce the cause. All subsequent 8D objectives depend on the accurate diagnosis of the root cause. Therefore, verification of the root cause is critical to the success of the 8D process. The flow for this analysis is shown in Figure 8.3.

Few things are more damaging to the problem-solving process than assigning blame. Blaming leads to defensiveness, and facts are then obscured or kept hidden. Misinformation is often generated as a defensive measure. Therefore, *avoid all blame and concentrate on the process!*

D4: Guidelines

As mentioned earlier, the fundamental purpose of D4 is to determine the root cause and escape point of the problem. Therefore, the focus is to identify both and verify them as the correct ones. Once identified, they are fixed at D6. This step identifies and evaluates the correct process of problem solving using the information from D2.

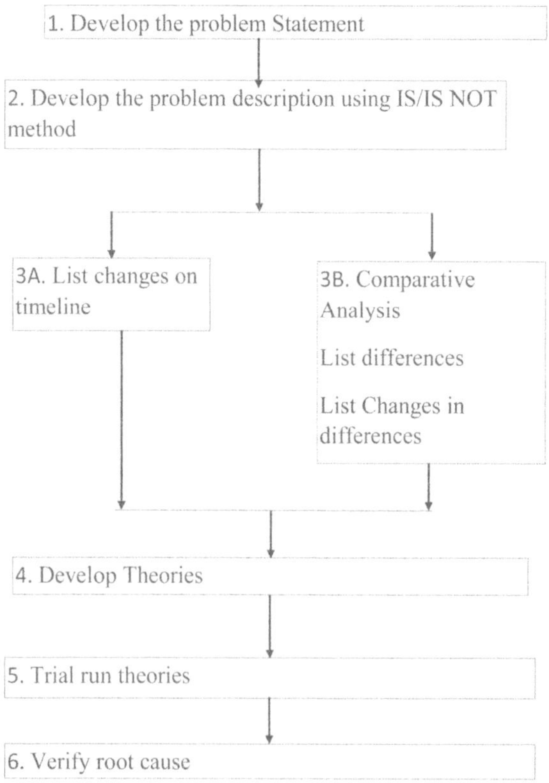

FIGURE 8.3 Determine root cause of the problem.

The basic guidelines begin with the appropriate definition of the root cause: The single verified reason that accounts for the problem and is followed by the escape point, which is: The earliest location in the process, closest to the root cause, where the problem should have been detected but was not. Other guidelines may deal with:

- Verification, which helps to make certain that PCAs are directed at the root cause and escape point. Time, money, effort, and resources are not wasted on false causes.
- Band-aids, which can mask information needed to find the root cause. Look out for band-aid fixes.
- The true root cause. Usually, the root cause is one change that caused the problem. If previous problems were never fixed at the root cause level, then the problem may be the result of more than one change.
- Deductive reasoning (first) to identify the possible causes. If this method does not work (due to missing information), then use the cause-and-effect diagrams and team members' experience to pursue other possible causes.
- Appropriate and applicable facts. You cannot identify the root cause without facts. The IS/IS NOT data must be correct, and the process flow diagram must be correct and up to date.
- All pertinent facts that are directly related to the root cause. The root cause should explain all known facts about the problem. Unexplained facts often indicate the presence of another root cause that creates a similar problem. Ideally, you should be able to make the problem come and go to prove you have identified the root cause.
- New surprises. When the root cause is uncovered, other troubles (which went unnoticed) are sometimes made visible. These other troubles create the need for an improved ICA in D3 and/or a return to D2.

D4: Evaluating Questions

When the team feels comfortable of having completed the D4 step, it is a good practice to review the discussion with at least the following basic questions:

- Has the factual information in the problem description been updated?
- What sources of information have been used to develop the potential root-cause list?
- Is there a root cause (a single verified reason that accounts for the problem)?
- What factor(s) changed to create this problem? What data are available that indicates any problem in the manufacturing or design process?
- How did we verify this root cause?
- Does this root cause explain all the facts compiled at D2?
- Has the root-cause analysis gone far enough (do we need to know why this root cause happened)?
- Is there more than one potential root cause?
- Does each item on the potential root-cause list account for all known data? Has each item been verified (used to make the defect come and go)?
- How did you determine assignment of the percent contribution?

- Combined, do the items on the potential root-cause list account for 100% of the problem (is the desired performance level achievable)?
- If the level is achievable, has the team considered and reviewed with the champion the benefit of developing a separate problem description (and, by definition, separate 8D) for each potential root cause?
- If the level is not achievable, has the team considered and reviewed with the champion the benefit of alternate problem-solving approaches?
- Do approaches independent of an 8D?
- Do approaches as a complement to an 8D?
- Does a control system exist to detect the problem?
- Has the current control system been identified? Does this control system represent a change from the original design?
- Has it been verified that the control system is capable of detecting the problem?
- Is the identified control point closest to the root cause/potential root cause?
- Is there a need to improve the control system?

A Generic Simple Root-Cause Analysis

Root-cause analysis is based on a questioning process focusing on the what, where; when, how big, and answering the following concerns about the cause using the IS/IS NOT method, identify differences, identify changes, and test the outcome. Figure 8.3 shows the flow of this analysis.

Determination of the Root Cause

In cases where "should" and "actual" were once the same but now are different, the root cause will be a change of some type. The definition of root cause is the single verified reason that accounts for the problem.

Two deductive approaches are available to lead the problem solver to the root cause. Both approaches are based on a thorough defect profile built around the problem statement. Both of these deductive approaches are a series of questions that yield answers to which another question is applied. The result is a steady reduction in the number of possible causes to be investigated.

For comparison, the steps of the two approaches are shown side-by-side in Table 8.2, having been anchored by the scope (problem statement and problem description). The two approaches are identical with one exception. Method B has an addition inserted in the third step.

TABLE 8.2
A Typical Comparison of Determining Root Cause Using Two Approaches

Approach A	Approach B
1. Establish a problem statement.	1. Establish a problem statement.
2. Develop a problem description.	2. Develop a problem description.
3. List ALL known changes on Timeline.	3a. List differences.
4. Develop theories based upon the changes.	3b. List changes in differences (only).
5. Trial run the theories.	4. Develop theories based upon the changes.
6. Verify the root cause/potential root cause.	5. Trial run the theories.
	6. Verify the root cause/potential root cause.

Method A is best used when all changes are known. On the other hand, Method B, with the additional step, eliminates consideration of changes common to both the IS and IS NOT listed in the problem description. Furthermore, Method B provides a hint where to investigate for hidden changes.

During either Method A or B, the problem solver may have to collect missing information. The problem solver can use probing questions to gather information from people who do the work (operators) and are more familiar with the details of the problem. Frequently, these people have essential information that they may not recognize as significant.

Test each theory against each discrete set of IS/IS NOTs listed in the Problem Description (step 2) for every element of "What," "Where," "When," and "How Big." Specifically, ask: "Does this 'change-how' theory completely explain both the IS and IS NOT?" With this question, the team is asking: "If this theory is the cause of the effect, do the factual data in the Problem Description explain fully why the effect manifests itself in the IS dimension but never manifests itself in the IS NOT dimension?" There are three possible outcomes:

1. *Explain fully*: If the data explain fully why the effect manifests itself in the IS dimension but never manifests itself in the IS NOT dimension, then a plus (+) symbol should be entered in the Test Matrix for that element. This theory fully explains this element.
2. *Cannot explain*: If the data cannot explain why the effect manifests itself in the IS dimension and/or never manifests itself in the IS NOT dimension, then a minus (–) symbol should be entered in the Test Matrix for that element. This theory definitely does not explain this element, and therefore cannot be the cause.
3. *Insufficient data*: If the theory could explain the effect but there are insufficient data to explain fully why the effect manifests itself in the IS dimension and/or never manifests itself in the IS NOT dimension, then a question mark (?) symbol should be entered in the Test Matrix for that element and a comment added at that point to indicate the further data collection or analysis that is required. This theory could explain this element, but more data or analysis is required. If, in fact, this is the situation, the team needs to:
   - Repeat the test question against each pair of IS/IS NOTs until a minus symbol is encountered for a theory, when it should be discounted and a further theory examined. The use of the minus (–) symbol is to exclude non-feasible theories from further consideration.
   - If only one theory passes through the Trial Run (with all pluses or a combination of pluses and question marks), then proceed to step 6 to verify the root cause included in this theory.
   - In practice, particularly where multiple potential root causes are considered, more than one theory may pass the Trial Run with a combination of pluses and question marks. In those cases, and where it is feasible and practical to do so, collect and analyze the missing data,

and re-examine the theory to resolve the question marks (?'s) to either pluses (+'s) or minuses (–'s) and proceed as above. If it is not feasible or practical to collect and analyze the missing data, proceed to step 6, starting with the theory with the most pluses (+'s).

Verify Root Cause

Verification of the possible causes can only be done in the real world, where the defect is occurring. Root cause verification cannot be done on a problem-solving form. Verification can be done in two ways – passive or active:

1. Passive verification is done by observation. With passive verification, you look for the presence of the root cause/potential root cause without changing anything. If you cannot prove the presence of the root cause/potential root cause variable, then chances are great that this possible cause is not the root cause.
2. Active verification is a process where the problem solver seeks to make the defect come and go using the variable thought to be the root cause/potential root cause. Both coming and going are important tests to confirm the root cause/potential root cause.

Helpful Supplemental Tools

The two deductive approaches to determine root cause are based on the availability of all relevant facts. They also presume to contain "myths" or information that is not fact. Sometimes, despite the best efforts of asking the right questions of the right people, the needed facts cannot be found. Supplements to the two deductive approaches include, but are not limited to:

- Statistical process control data.
- Process flow diagrams.
- Process flow cause-and-effect diagram.
- Possible causes listed on a cause-and-effect diagram.
- A subject matter expert's experience.
- Part analysis of the defect.
- Distinction analysis.
- Group brainstorming.
- Force field analysis.
- 5 Whys analysis.
- 3 × 5 whys analysis.

Each of the above exhibits both benefits and liabilities when used with a purely deductive method. For more tools, see Appendices B and C.

D5: Choose and Verify Permanent Corrective Actions (PCAs) for Root Cause and Escape Point

The purpose of D5 is to select the best PCA to remove the root cause. Also, select the best permanent corrective action to eliminate the escape point. Finally, another salient purpose of the D5 is to verify that both decisions will be successful when implemented without causing undesirable effects.

The 8D process is applied to fix a problem at the root-cause level and prevent its recurrence. In change-induced conditions, the root cause(s) is introduced by a change. In an overwhelming number of cases, the change is introduced by someone's decision. At the time of the decision, the problem is not anticipated. The D5 objective is designed to avoid a similar sequence of events. The historical function of D5 is to make the best decision on how to remove the root cause variable that created the defect. Because decisions potentially lead to other effects, the decision needs to be thoroughly examined before its implementation at D6. Customers and managers must be involved.

The decision is first examined for both its benefits and its risks. Then, the decision is tested (verified) to ensure it will be successful when implemented at D6. The D5 objective ends with a commitment to full-scale implementation of this decision at D6. The decision-making process depends on the experience of the decision makers and the criteria that are applied. This is especially true during the risk analysis phase of the decision-making process. Experience is also critical to the verification segment of D5. Experience guides how verification should be conducted and how much data is appropriate. [Of special note here is worth mentioning that the champion's commitment is of paramount importance and essential. The criteria on which the decision is based are typically a synthesis of the champion's criteria and the detailed experience of the team members.]

D5: Process Guidelines

As mentioned earlier, the fundamental purpose of D5 is to select the best corrective action(s) to eliminate the root cause and escape point of the problem. Therefore, more than one permanent corrective action may be required to resolve 100% of the problem. The team should remember that the D5 decision is not implemented until D6. In addition, the team should consider having risk analysis conducted by those who must implement the corrective action. Some implied actions of D5 are:

- Check the composition of the team. Does the team have the right experience to make the decision(s)?
- Decide what is the best permanent corrective action?
- Evaluate each choice from both its benefits and liabilities.
- Test, if necessary, the decision to determine that the choice (action) will, in fact, work.
- If this PCA cannot be implemented quickly, will the D3 ICA last long enough?
- Get approval (through the champion) to proceed with the choice. Implement the decision at D6.

In addition to these implied actions, it is necessary that it must be recognized that:

1. Sometimes the best PCAs cannot be implemented due to cost, limited resources, *etc*. Sometimes the root cause cannot be eliminated. In these cases, the best decision is to continue using the D3 ICA. This measure is a compromise, but it is sometimes the only resolution. This should be considered an exception for D5, not the norm. Even if this action is appropriate, D6, D7, and D8 should still be completed.
2. Sometimes the best PCA will require time before it can be implemented. In this case, the D3 ICA will continue until replaced by the D6 PCA. The 8D process permits a redesign of the ICA. This redesigned ICA might be more effective and less costly than the initial D3 ICA.
3. Sometimes during D5 verification, the presence/evidence of another root cause that was not discovered during D4 may be uncovered. This is indicated when the verified PCA does not remove 100% of the unwanted effect.

D5: Evaluating Questions

When the team feels comfortable of having completed the D5 step, it is a good practice to review the discussion with at least the following basic questions:

- What criteria have been established for choosing a PCA for the root cause and escape point? Does the Champion agree with these criteria?
- Is a service action required as part of the PCA?
- What choices have been considered for the PCAs?
- Have we overlooked better choices?
- What features and benefits would the perfect choice offer? How can we preserve these benefits?
- Do we have the right experience on this team to make this decision?
- What risks are associated with this decision, and how should they be managed?
- Does the champion concur with the PCA selections?
- What evidence (proof) do we have that this will resolve the concern at the root-cause level?
- Did our verification approaches make allowances for variations in the frequency (or patterns) created by the cause?
- Which variables did we measure during the verification step? Do these indicators constitute sound verification?
- What are the possibilities that this choice, once implemented, will create other troubles?
- Can our customer live with this resolution?
- Will our containment continue to be effective until our choice can be implemented?
- What resources will be required for PCA implementation? Do we have these resources?
- What departments will need to be involved in the planning and implementation of this decision?
- Have actions been considered that will improve the ICA prior to PCA implementation?

D6: Implement and Validate Permanent Corrective Actions (PCAs)

The purpose of D6 is to plan and implement selected PCAs, remove the ICA, and monitor the long-term results. In D5, the PCA was chosen and verified. In D6, the PCA is implemented, and the problem is corrected. In D6, the team is responsible for detailed planning, implementation, and evaluation.

One of the major responsibilities of the champion's authority is the requirement to authorize making changes to resolve the problem. Generally, even relatively minor changes will require cross-functional participation. For example, changing the length of a purchased bolt impacts the documentation of purchasing, quality assurance, material control, inventory control, engineering, and production departments. In this simple example, the champion might be the production department manager, but the production department still requires communication and close coordination.

Cross-functional involvement at D6 also means cross-functional planning to implement the D6 correction. During D6, team membership may expand temporarily to facilitate complete planning. This planning may or may not require a group meeting on the subject, but some form of contact with other departments is essential for successful implementation. At a minimum, the existing team must identify who (in which department) needs to be involved in planning the implementation of D6.

Removing the ICA

Embedded in D6 is removal of the D3 ICA. The ICA is no longer needed because the PCA removes the root cause variable. Continuing the ICA wastes important resources. Further, the ICA has masked the problem, so validation of the D6 PCA would be questionable.

D6 planning must account for when to terminate the D3 ICA and how to validate the results of the D6 PCA. Technically, this validation is continuous and ongoing, but validation is likely to be more frequent at first, tapering off over time. In simple terms, the events of D6 could be expressed by the model in Figure 8.4.

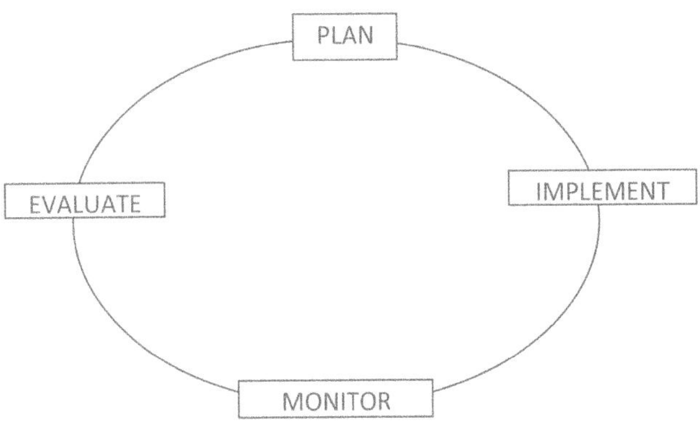

FIGURE 8.4 The validation model of removing the ICA.

The team members' and champion's active involvement occurs at each event, although the activities of each tend to be different.

D6: Process Guidelines

The critical task of updating all documents, procedures, process sheets, *etc.*, is too frequently overlooked. The D6 PCA fixes a problem by making changes. If the PCA change is not documented in the normal systems, practices, and procedures, then additional corrections will likely be necessary. After all, often a problem recurs when someone reads an obsolete instruction. Related problems also occur, such as ordering obsolete parts and being unable to order needed parts that are not recorded on the bill of materials in the material control system.

As we mentioned before, the purpose of D6 is to implement the decision made at D5. This implementation, however, is done via the management systems. They are not bypassed. The implied actions of D6 include:

- Check the composition of the team. Is the needed experience on the team?
- Verify that the decision needs to be planned, implemented, monitored, and evaluated for effectiveness.
- Update any changes implemented during D6. All updates must be documented (formally) in process flow sheets, procedures, practices, instructions, standards, spec sheets, FMEAs, control plans, *etc.*, in the departments that maintain such records or documents. The updates need to have at least three minimum items for any action plan: (a) What, (b) Who, and (c) When due.

Furthermore, it should be recognized that:

- D6 requires detailed, hands-on people to be involved in planning D6 actions.
- The 8D process does NOT supersede an organization's procedures to implement process changes or customer approval processes.
- Some changes may require approval of your customer(s) before the changes can be made.
- No changes should be made without the champion's direct approval.
- Some people may require training before you can implement the D6 change because their procedures may be affected.
- During D6, the D3 action should be stopped because the ICA is no longer necessary. Continued D3 actions may mask a portion of the defect that is still present.
- D6 is not completed until the PCA is validated (the customer no longer experiences the problem). Prudence dictates that validation should occur at some point before the customer can experience the problem.
- Long-term monitoring is appropriate after the D6 PCA is implemented.
- Failure to document the correction created by the D6 PCA will eventually reintroduce the root cause. Failure to document the PCA typically results in overstocking obsolete parts and stockouts of needed parts.
- Consider adopting the following statement as a normal D6: Flawless implementation of the PCA, where things work right the first time.

D6: Evaluating Questions

When the team feels comfortable of having completed the D6 step, it is a good practice to review the discussion with at least the following basic questions:

- What departments are needed to implement the PCAs?
- Are representatives of those departments on our team to plan and implement their parts?
- What customer and/or supplier involvement is needed?
- Who will do the planning for the customer? For the supplier?
- Has an action plan been defined (responsibilities assigned, timing established, required support determined)?
- Do we have the necessary resources to implement this plan? What is needed?
- At what point(s) is this plan vulnerable to things-gone-wrong? What can be done to prevent these?
- What are appropriate contingent actions?
- What will trigger our contingent actions?
- How are we monitoring completion of the plan?
- When will the ICA be removed?
- How will we communicate this plan to those who have a need to know? What training will be required?
- What measurable(s) will be used to validate the outcome of the PCAs (both short-term and long-term)?
- Has the ICA been discontinued?
- Has the unwanted effect been totally eliminated?
- How can we conclusively prove this?
- How are we continuing to monitor long-term results? What is the measurable? Is this the best way to prove the root cause is eliminated?
- How have we confirmed the findings with the customer?
- Have all systems, practices, procedures, documents, *etc.* been updated? Do they accurately reflect what we want to be done from here on?

D7: Prevent Recurrence

The purpose of D7 is to modify the necessary systems, including policies, practices, and procedures, to prevent recurrence of this problem and similar ones. It also helps the management to infuse the changes into the organization as applicable. Even though the D7 is frequently cited as perhaps the most important step of the 8D process, it is often skipped and overlooked because of the fact that "the" temporary fix is working. During D7, the root cause of the D4 is fixed. However, behind every change that creates the need to do an 8D is at least one system, practice, or procedure that introduced the change. If the system, practice, or procedure remains the same, the same or a similar problem-causing change will be reintroduced. This can and should be prevented.

So, even though D7 is a very substantial step in the process, many practitioners do not complete it for many reasons. We will identify some of the reasons, and hopefully, making the reader aware of these reasons will help the organization to develop a program to combat them.

- D7 is not even attempted. After the problem is corrected, other priorities get more attention. Infusion is forgotten.
- D7 develops into a blame session. Rather than focus objectively on what happened, the attention drifts to "who" did/did not do something. Defensiveness sets in, and objective review stops.
- The 8D process stopped at D3 containment. The root cause was never found. Prevention of the root cause is impossible because it remains unknown.
- The 8D process was never done. Someone penciled in a form that became an 8D Report.
- Politics: Fixing systems is a low priority because it requires risking (forcing) a change.
- Fear: People perceive/know that spotlighting defective/inadequate system practices and procedures will result in reprisal.
- Not enough structure/authority: The champion is too low in the organization's power hierarchy to implement necessary system revisions. Further, this champion is unable to solicit the involvement of those who have the authority.
- Those who own the system(s) do not have to live with the consequences: The organization is unwilling to create consequences at the proper points of systems management.
- Systems maintenance and management is a low priority. Activity is more important than prevention. Process control is not understood.

Some of these issues obviously exceed the scope of one champion's authority. However, they can be corrected. Some take a lot of effort and time. The first step of any D7 objective is to accurately identify (at least) the cause of the system, process, *etc.*, which introduced the root cause. This much the champion can prompt with the team's understanding of the first six objectives. D7 often requires bringing all former team members together to answer two questions:

1. What went wrong that introduced the root cause in the first place?
2. What needs to be fixed, modified, or reinforced to prevent this from happening again?

On the other hand, we must recognize that:

- D7 also implies some opportunities. These include the chance to suggest what a good system should look like and to identify places where a defect is about to be created.
- The improvements that may follow beyond the process of original concern depend heavily on the rank, authority, and power of the champion. Generally, there are three paths that may be followed after the systemic issues are identified:
    1. The champion uses his/her authority to fix the system, practice, or procedure.
    2. The champion carries the recommendation for systemic changes to those who have the authority to change the system, *etc.*

3. The need for systemic changes is acknowledged, but the organization chooses to actively monitor the day-to-day operation of the system. This is done to better understand the whole system, rather than overreact to a single incident. Other 8Ds might also be reviewed for their recommendations and observations of the system's shortcomings.

Of course, we must also not forget that the 8D process asks for input at D7. The champion retains 51% of the vote. D7 does not dictate that the champion implement every recommendation. However, the champion works against his/her own best interest if ideas are discounted without serious consideration.

D7: Process Guidelines

As already mentioned, the purpose of D7 is to prevent the recurrence of this kind of root cause and escape point, as well as to infuse the solution to the problem in the organization as appropriate. In addition to the fundamental purpose of the D7, the following items should be recognized and considered for review, as well as discussed as needed:

- Approximately 95% of all root causes are introduced because of a procedure, practice, instruction, or management system malfunction. These systems may have been ignored, they may have never worked initially, or they may be too cumbersome or obsolete. If the system is not fixed, it will create a similar kind of problem and/or another root cause.
- Typically, the team is best able to identify the details of the systemic problem.
- The team is chartered with both the responsibility and the authority (empowered) to provide feedback to management (through the champion) as to which systems contributed to the root cause and escape point.
- The team members will draw from all the knowledge and experience gained from the first six steps. The champion may have to take this information to other areas within management to implement systemic recommendations because he/she may not have full authority to make changes as recommended.

Not all team-identified prevent actions and systemic recommendations will be the best and should not be implemented. But in fairness to the team, reasons why these will not be implemented should be communicated. It is very unfortunate that this is hardly done by practitioners.

If the team has the skill, experience, and knowledge, the champion can ask members to design and implement their prevent actions. If the team does not have the skill, *etc.*, then the champion will request that other areas of the organization design and implement the change(s).

- D7 works best when the people who use the systems have had a hand in the changes. Think simple, easy, and quick in system redesign. Usually, adding more words or instructions adds to confusion as opposed to removing confusion. Practice the keep it simple, Stephanie (KISS) principle.
- D7 is best achieved by a face-to-face review between the champion and the team.
- The team members should recognize that the champion may have to reflect upon their recommendations before being able to suggest or approve changes.
- The team should follow the format of observation first, recommendation second, in the presentation to the champion.
- The team and champion should avoid creating a blame session.
- Sometimes the systems that contributed to the root cause are so big that not one individual in an organization or company owns all parts of that system. For example: Cost accounting systems, order entry systems, *etc.*, may require a separate task force by the company to redesign them.
- The champion and team require cooperation and objectivity from each other and from the rest of the organization to make D7 a success.
- The champion's reputation and actions often precede him/her in the D7 face-to-face phase. Most teams will only put as much effort into the D7 steps as they believe the champion will support.
- It is not uncommon for team members to feel threatened at D7. Therefore, the champion must communicate, before D7, his/her interest in receiving legitimate feedback on systems, practices, and procedures. The champion must be careful not to create unrealistic expectations.
- The team needs to realize that champions often do not know about every detail and consequence of all the systems, practices, and procedures they are expected to manage. This is the team's opportunity to objectively (not emotionally) point out flaws in day-to-day operations in procedures.
- The leader may have to point out that most D7 actions make life easier in the workplace, not more difficult.
- Use of the Repeated Why method is an effective technique during D7. However, it can also be very frustrating for those trying to answer it. Use it carefully.

D7: Evaluating Questions

When the team feels comfortable of having completed the D7 step, it is a good practice to review the discussion with at least the following basic questions:

- How and where did this problem enter our process?
- Why did the problem occur there, and how did it escape? (Why was it not detected?)
- Did confusion or lack of knowledge contribute to the creation of this root cause? Did it contribute to the escape?
- What policies, methods, procedures, and/or systems allowed this problem to occur and escape?

- Did you uncover Band-Aid© fixes in our processes? Where? What are they compensating for?
- Have the affected parties been identified?
- What needs to be done differently to prevent recurrence of the root cause? Of the escape?
- Is a service action required as part of the prevent actions?
- What evidence exists that indicates the need for a process improvement approach (*i.e.*, Focused improvement, re-engineering)
- Who is best able to design improvements in any of the systems, policies, methods, and/or procedures that resulted in this root cause and escape?
- What is the best way to perform a trial run with these improvements?
- What practices need standardization?
- What plans have been written to coordinate prevent actions and standardize the practices - who, what, and when?
- Does the champion concur with the identified prevent actions and plans?
- How will these new practices be communicated to those affected by the change?
- Have the practices been standardized?
- What progress check points have been defined to assess system improvements?
- What management policy, system, or procedure allowed this problem to occur or escape?
- Are these practices beyond the scope of the current champion?
- Who has responsibility for these practices?
- Does the current champion agree with the systemic prevent recommendations of the team?
- What data has been submitted to the organization's Lessons Learned database?

D8: Recognize Team and Individual Contributions

The purpose of D8 is to complete the team experience, sincerely recognize both team and individual contributions, and celebrate. Therefore, the D8 is the closure of the team effort to resolve a problem. Because of this closure, D8 is a very important step even though many practitioners do not do it. At the very least, it is a clear message for everyone to return to their full-time job. D8 says to everyone that the task team is now disbanded and no further study of this concern is justified. More importantly, D8 is one way for the organization to provide feedback to those who did the work (D1 through D7). However, how the message is conveyed to the team is nearly as important as the message itself. It should be fit, focused, and timely. Recognition at D8 includes noting the team's contribution and efforts at all earlier steps. If recognition is omitted, team members interpret that their contributions were not valued. If the 8D process was not implemented using people, D8 could be left out.

To be sure, volumes have been written on recognition – the need for it, how to do it, and how to do it well. This brief section is not an attempt to repeat or condense that. A more complete study on the subject is well advised, but a few highlights are helpful.

- Encourages a repeat of the positive (wanted) behavior.
- Works best when it is perceived to be sincere.
- Is expected when extra effort has been expended.
- Does not have to be expensive in dollars, but does require investment in time; time to find out exactly what was done, time to express the recognition (personal time).
- Is situational.
- It must be on a timely basis.
- Works best when the method of recognition is unique, a bit of a surprise, and considered valuable. Value does not mean money. Often value means personalized, customized.
- Recharges a person's battery. It feels good. People tend to do things for which they are recognized.
- Several people can benefit from recognition: Team members, the team's leader, former team members who were released from the team, and the recorder and facilitator. Each provided important services and made contributions. Recognition is a way to encourage future participation. The more effectively recognition is used, the better the organization's results.

D8: Process Guidelines

As mentioned, the purpose of D8 is to recognize the contributions made to the company or department by the 8D team. There is a difference, however, between recognition and reward. Either or both can be used at D8. The responsibility for recognition belongs to the champion. However, legitimate recognition opportunities exist between team members, the leader and the team, the team and the champion, *etc.* Messages about organizational values will be communicated to the rest of the organization by the D8 action. Failure to conduct D8 will also send messages throughout the organization. Additional consideration and opportunity for either review or open discussion should be encouraged for the team participants. Some typical items may be the following:

- Care must be taken that the D8 action is commensurate with contribution of the 8D team.
- Care must be taken that the D8 action is perceived by the team members to be a legitimate and sincere acknowledgment of their efforts.
- It is recommended that variety and creativity be applied in identifying what the champion's D8 actions are. Additional guidelines for success include:
  - Fit
  - Focus
  - Timeliness

The action should fit the contribution of the team. Verbal recognition should focus (accurately describe) the team's actions and contributions. The actions should also be completed in a timely manner. Make certain all members, both core and extended members, have the opportunity to participate. After all, D8 is an opportunity to encourage certain behaviors and actions. If the team's contributions at steps D5, D6, and D7 are not acknowledged, then the team would rightly presume that these actions do not matter.

If D8 actions are consistently the same and predictable, the 8D process will be perceived as another lip service program. Generally speaking, people look for some legitimate, personal involvement of the champion for the D8 step to be interpreted as sincere recognition.

D8: Evaluating Questions

When the team feels comfortable of having completed the D8 step, it is a good practice to review the discussion with at least the following basic questions:

- Has the 8D report been updated and published?
- Is the 8D paperwork completed and have other members of the organization (who have a need to know) and the customer been informed of the status of this 8D?
- Is the 8D report and its attachments retained in our historical file system?
- Is there a complete list of all team members, current and past?
- Were there significant contributions by individual team members? What were they?
- Are there opportunities to provide recognition from leader to team, team member to team member, team to leader, team to champion?
- What are some different ways to communicate the recognition message?
- Are there any non-team members whose contributions to the 8D process justify inclusion at recognition time?
- Are all current and past team members being recognized?
- Do the results achieved by the team warrant some publicity (*e.g.*, company newsletter)?
- Does the recognition satisfy the fit, focus, and timely criteria?
- D8 is intended to be positive. What are the chances that it might backfire and turn into a negative?
- What have you learned as individuals and as a team? About yourselves? About problem solving? About teamwork?
- How did the organization benefit from the completion of this 8D process?
- Review each 8D objective. What was done well? What was not done so well – could it be improved? If yes, how?
- What sort of things should be repeated if conditions bring them together, escape on another 8D?
- Are there changes to the business practices that should be considered, based on the learning in this 8D?
- Has all unfinished business now been finalized?
- Has each person had an opportunity to express appreciation to other team members?
- Has each person made a final statement?
- Have you celebrated appropriately? How?

8D: Summary Evaluation

When the 8D is complete, it is a good practice to review the entire process. A good review summary for any 8D process is the following. Please notice the tools used for each step. These are obviously simple but common and very powerful. There are many more available – see Appendices B and C – and the reader may want to use others as appropriate.

1. **Prepare for the 8D Process**
   a. The symptom has not been defined.
   b. The symptom has not been quantified.
   c. The need for an ERA has not been evaluated.
   d. The ERA has not been validated.
   Typical Quality Tools: Trend Chart, Pareto Chart, Paynter Chart.
2. **Establish the Team**
   a. The Champion has not been named.
   b. The sub-supplier(s) is/are not on the team.
   c. The customer is not on the team.
   d. Other pertinent personnel are not on the team.
   e. No indication of a team Brainstorming activity.
3. **Describe the Problem**
   a. The problem is not stated in the customer's terms.
   b. The start/end date of the problem is not listed.
   c. The percentage defective is not listed.
   d. The frequency of the problem is not listed.
   e. The actual reported measurement is not listed.
   f. The requirements (specifications) are not listed.
   Typical Quality Tools: Is/Is Not Analysis, 5 Whys, 3 × 5 Whys, Pareto Chart, Process Flow Diagram, Cause and Effect Diagram.
4. **Develop the Interim Containment Action**
   a. All stock locations were not purged of suspect stock.
   b. Supplier's site.
   c. Customer's site.
   d. In transit.
   e. Warehouse.
   f. Next customer's site.
   g. Sub-Supplier's site.
   h. Other.
   i. The percentage of effectiveness of the containment is not listed.
   j. The percentage of effectiveness of the containment is unsatisfactory.
   k. The dates of the containment actions are not listed.
   l. Sort results do not list the ____ # Sorted, ____ # Found, ____ # Defective.
   m. The method of sort is.
   n. Not stated.
   o. Unsatisfactory.

p. Containment action is compromised solely by an adult.
q. No short-term corrective actions are listed.

Typical Quality tools: Paynter Chart, SPC Data, FMEA.

5. **Define and Verify Root Cause and Escape Point**
   - No Fishbone Analysis.
   - No Is/Is Not Analysis.
   - Unacceptable root cause.
   - Operator error.
   - Set-up piece.
   - Stated root cause is NOT a root cause (either a symptom or effect).
   - The percent contribution is each root cause is not listed.
   - Failure to recognize PROCESS root cause.
   - Failure to recognize SYSTEM root cause.
   - No verification of root cause.
   - Process flow diagram missing when the nature of the problem clearly requires it.
   - The "Five Whys" have not been explored.
   - An escape point has not been identified.
   - Whether or not a control system exists to detect the problem has not been identified.
   - The capability of the control system has not been evaluated.

   Typical Quality Tools: Cause and Effect Diagram, FMEA, Is/Is Not Analysis, 5 Whys, 3×5 Whys, DOE, Robust Design.
6. **Choose and Verify Permanent Corrective Actions for Root Cause and Escape Point**
   - The dates and actions are not indicated.
   - The stated actions do not cover all PROCESS root causes.
   - The stated actions do not cover all SYSTEM root causes.
   - The actions do not adequately address the timing of the problem/condition/countermeasure requirements.
   - The actions do not adequately address the magnitude of the problem.

   Typical Quality Tools: Paynter Chart, Cause and Effect Diagram, DVP&R, 5 Whys, 3×5 Whys, FMEA, Robust Design.
7. **Implement and Verify Permanent Corrective Actions**
   - The percentage effectiveness of the corrective actions has not been listed.
   - The Paynter Chart, or components of it, used to verify the effectiveness, is missing.
   - Throughput.
   - Return sales quantity.
   - Scrap.
   - Return sales sort results.
   - Dock audit.
   - Warranty.
   - Incoming rejects (IRs).
   - The actions listed are DETECTION actions.

   Typical Quality Tools: FMEA, Paynter Chart, Is/Is Not Analysis, 5 Whys, 3×5 Whys SPC, PPAP.

8. **Prevent Recurrence**
   - The PFMEA was not reviewed.
   - The PFMEA was not adequately revised.
   - Control Plans were not revised.
   - Process Sheets were not revised.
   - Prevention Action is composed solely of an audit.
   - Containment action is comprised of an added in-process inspection.

   Typical Quality tools: FMEA, Control Plans, ISO/QS Procedures, Fault Tree analysis, Process Flow Diagram.
9. **Recognize Team and Individual Contributions**
   - The 8D Report has not been updated.
   - All parties who need to know have not been advised of the status of the 8D.
   - All unfinished business has not been finalized.

8D: Success Factors

To be sure, the 8D is a very powerful methodology to resolve problems. However, its success depends on the commitment of the management of the organization. This means in no uncertain terms that:

1. *Management MUST create the right climate*: This means that they must:
   - Display patience.
   - Provide practice time.
   - Demonstrate interest – even when failure is experienced.
   - State expectations clearly.
2. *Management's recognition of need to prioritize problems*. This means that they must:
   - Continuously improve the system.
   - Expectations being realistic in every respect.
   - Provide the right resources.
   - Communicate the significance of the problem with the team and everyone concerned with the specific problem.
3. *Management's active support for problem-solving process*. This means that they have to demonstrate interest by:
   - Giving authority and responsibility to the appropriate individuals to carry out the assigned task.
   - Always reviewing progress of the project and asking coaching questions.
   - Making sure that the process defined is adhered to.
   - Being an implementation advocate.
   - Providing a system to retain the "gained" knowledge.
   - Providing the appropriate resources.

4. *Management's commitment for a system to carry out corrective actions.* This means that management must practice empowerment with both authority and responsibility for those individuals who must carry out the task assigned.
5. *Management must encourage prevention of problems.* To correct a problem is Ok, but it is after the fact. To prevent a problem needs a commitment to advance planning so that problems do not occur.

9 Common Methodologies for Customer Service

POKA-YOKE

OVERVIEW

The term *poka-yoke* was applied by Shigeo Shingo in the 1960s to industrial processes designed to prevent human errors (Robinson 1997). Shingo redesigned a process in which factory workers, while assembling a small switch, would often forget to insert the required spring under one of the switch buttons. In the redesigned process, the worker would perform the task in two steps: First, preparing the two required springs and placing them in a placeholder, then inserting the springs from the placeholder into the switch. When a spring remained in the placeholder, the workers knew that they had forgotten to insert it and could correct the mistake effortlessly (Shingo 1987).

Shingo distinguished between the concepts of inevitable human mistakes and defects in production. Defects occur when the mistakes are allowed to reach the customer. The aim of poka-yoke is to design the process so that mistakes can be detected and corrected immediately, eliminating defects at the source.

HISTORICAL PERSPECTIVE

Poka-yoke can be implemented at any step of a manufacturing process where something can go wrong or an error can be made (Quality Portal 2013). For example, a jig that holds pieces for processing might be modified to only allow pieces to be held in the correct orientation, or a digital counter might track the number of spot welds on each piece to ensure that the worker executes the correct number of welds (Nikkan 1988).

Shingo recognized three types of poka-yoke for detecting and preventing errors in a mass production system (Shingo and Dillon 1989; Quality Portal 2013). They are:

1. The *contact* method, which identifies product defects by testing the product's shape, size, color, or other physical attributes.
2. The *fixed-value* (or *constant number*) method, which alerts the operator if a certain number of movements are not made.
3. The *motion-step* (or *sequence*) method, which determines whether the prescribed steps of the process have been followed.

One can see that from the three options, either the operator is alerted when a mistake is about to be made, or the poka-yoke device actually prevents the mistake from being made. In Shingo's lexicon, the former implementation would be called

DOI: 10.1201/9781003650614-9

a *warning* poka-yoke, while the latter would be referred to as a *control* poka-yoke (Shingo and Dillon 1989).

Shingo argued that errors are inevitable in any manufacturing process, but that if appropriate poka-yokes are implemented, then mistakes can be caught quickly and prevented from resulting in defects. By eliminating defects at the source, the cost of mistakes within a company is reduced. This is because defects are a form of waste. Waste is not value added and therefore, if present, productivity and profitability are decreased. A methodic approach to build up poka-yoke countermeasures has been proposed by the applied problem solving (APS) methodology (Fantin, 2014), which consists of a three-step analysis of the risks to be managed:

1. Identification of the need.
2. Identification of possible mistakes.
3. Management of mistakes before satisfying the need.

This approach can be used to emphasize the technical aspect of finding effective solutions during brainstorming sessions.

THE STRATEGY OF MISTAKE PROOFING

The strategy of mistake proofing is to establish a *team approach* to "Mistake Proof Systems" that will focus on both *internal* and *external* customer concerns. This will include quality indicators such as Online Inspection and Probe Studies. Appendix C provides several items of interest in the area of signals, forms, and rationale of defects and mistakes. A strategy of mistake proofing involves the following:

- Concentrating on the things that can be changed rather than on the things that are perceived as having to be changed to improve process performance.
- Developing the training required to prepare team members.
- Involving all the appropriate people in the Mistake Proof Systems Process.
- Tracking quality improvements using in-plant and external data collection systems (before/after data).
- Developing a "core team" to administer the Mistake Proof Systems Process. This core team will be responsible for tracking the status of the Mistake Proof Systems throughout the implementation stages.
- Creating a communication system for keeping Plant Management, Local Union Committee, and the Joint Quality Committee informed of all progress, as applicable.
- Developing a process for sharing the information with all other departments and/or plants, as applicable.
- Establishing the "mission statement" for each team and "objectives" that will identify the philosophy of Mistake Proof Systems as a means to improve quality.
- Developing timing for completion of each phase of the process.
- Establishing cross-functional team involvement with your customer(s).

As any strategy must have a mission, so does the mistake-proofing approach. A typical mission statement may read: *To protect our customers by developing mistake proofing systems that will detect or eliminate defects while continuing to pursue variation reduction within the process.* Typical objectives to support the mission may be to:

- Become more aware of quality issues that affect the customer.
- Focus the efforts on eliminating these quality issues from the production process.
- Expose the conditions that cause mistakes.
- Understand "source investigation" and recognize its role in preventing defects.
- Understand the concepts and principles that "drive" mistake prevention.
- Recognize the three (3) functional levels of Mistake Proof Systems.
- Be knowledgeable of the relationships between Mistake Proof Systems devices and defects.
- Recognize the "key" Mistake Proof System devices.
- Share the Mistake Proof System knowledge with all other facilities within the organization.

Why must we have a strategy? Because many things can and often "go wrong" in our ever-changing and increasingly complex work environment. Opportunities for mistakes are plentiful and often lead to defective products. Defects are not only wasteful but also result in customer dissatisfaction if not detected before shipment. The philosophy behind Mistake Proof Systems suggests that if we are going to be competitive and remain competitive in a world market that

we cannot accept any number of defects (nonconformances) as satisfactory!

In essence, not even one defect can be tolerated. Mistake Proof Systems is a simple method for making this philosophy become a daily practice. Simple concepts and methods are used to accomplish this objective.

MISTAKE PROOF SYSTEMS

Humans tend to be forgetful and as a result - make mistakes. In a system where *"blame" is* practiced and people are held accountable for their mistakes and mistakes within the process, result in discouraging the worker and lower morale of the individual - but the problem continues and remains unsolved. To remedy this, the mistake proof system is implemented, which is a *technique for avoiding errors in the workplace.*

The concept of error-proof systems has been in existence for a long time, only we have not attempted to turn it into a formalized process. It has often been referred to as: "idiot proofing," "goof proofing," "fool proofing," *etc.* These often have a negative connotation that appears to attack the "intelligence" of the individual involved, and therefore are not used in today's work environment. For this reason, we have selected the terms *Mistake Proof Systems.* The idea behind error proof systems is: *to reduce the opportunity for human error by taking over tasks that are repetitive, or actions that depend solely upon memory or attention.* With this approach, we allow the workers to maintain dignity and self-esteem without the negative connotation that the individual is an "idiot," "goof," or "fool."

There are many reasons why mistakes do happen! However, almost all the mistakes can be prevented if we diligently expend the time and effort to identify the basic conditions that allow them to occur, such as: (a) when they happen, and (b) why they happen. When we perform a diligent study for a given defect, we can then determine what steps are needed to prevent these mistakes from recurring – *permanently.* The Mistake Proof System approach and the methods used give us an opportunity to prevent mistakes and errors from occurring.

In the case of human mistakes, we must understand the system that allows the individual to make the mistake and the conditions that an individual is operating under to do a particular task. In the following list we have identified several of the conditions that one may be prone to create a mistake and ultimately a defect.

1. *Forgetfulness*: There are times when we forget things, especially when we are not fully concentrating or focusing. An example that can result in serious consequences is the failure to lock out a piece of equipment or machine we are working on. To preclude this, precautionary measures can be taken: post-lockout instructions at every piece of equipment and/or machine, and have an on-going program to continuously alert operators of the danger.
2. *Mistakes of misunderstanding*: Jumping to conclusions before we are familiar with the situation often leads to mistakes. For example: Visual aids are often made by engineers who are thoroughly familiar with the operation or process. Since the aid is completely clear from their perspective, they may make the assumption (and often do) that the operator fully understands as well. This may not be true. To preclude this, we may:
   a. Test this hypothesis before we create an aid.
   b. Provide training/education.
   c. Standardize work methods and procedures.
3. *Identification mistakes*: Situations are often misjudged because we view them too quickly or from too far away to clearly see them. For example: Misreading the identification code on a component of a piece of equipment and replacing that component with the wrong part. To prevent these, we might:
   a. Improve legibility of the data/information.
   b. Provide training.
   c. Improve the environment (lighting).
   d. Reduce boredom of the job, thus increasing vigilance and attentiveness.
4. *Amateur errors*: Lack of experience often leads to mistakes. A newly hired worker will not know the sequence of operations to perform their task, and often, due to inadequate training, will perform those tasks incorrectly. To prevent amateur errors:
   a. Provide proper training.
   b. Utilize skill-building techniques prior to job assignment.
   c. Provide work standardization.

5. *Willful mistakes*: Willful errors result when we "choose" to ignore the rules. For example: Placing a rack of material outside the lines painted on the floor that clearly designates the proper location. The results can cause damage to the part, the material, or perhaps an unsafe work condition. To prevent this situation:
   a. Provide basic education and/or training for the task and safety.
   b. Require strict adherence to the rules.
6. *Inadvertent mistakes*: Sometimes we make mistakes without even being aware of them. A wrong part might be installed because the operator was daydreaming. To minimize this from occurring, we may:
   a. Standardize the work.
   b. Create appropriate discipline, if necessary.
7. *Slowness mistakes*: When our actions are slowed by delays in judgment, mistakes are often the result. For example: An operator unfamiliar with the operation of a forklift, might pull the wrong lever and drop the load. Methods to prevent this might be:
   a. Skill building.
   b. Work standardization.
8. *Lack of standards mistakes*: Mistakes will occur when there is a lack of suitable work standards or instructions that are not understood by the workers. For example: Two inspectors performing the same inspection may have different views on what constitutes a reject. To prevent this:
   a. Develop *operational definitions* of what the product is expected to be that are clearly understood by all.
   b. Proper training and education.
9. *Surprise mistakes*: When the function or operation of a piece of equipment suddenly or unexpectedly changes without warning, mistakes may occur. For example: Power tools that are used to supply specific torque to a fastener will malfunction if an inadequate oil supply is not maintained in the reservoir. Errors such as these can often be prevented by:
   a. Work standardization.
   b. Having a Total Productive Maintenance System in place. Typical basic items of concern should be in the areas of: (a) precision alignment, (b) imbalance, (c) lubrication, (d) appropriate bearing installation, and (e) visual inspection.
10. *Intentional mistakes*: Mistakes are sometimes made deliberately by some people. These fall in the category of sabotage. Disciplinary measures and basic education are the only deterrents to these types of mistakes.

ESSENTIALS OF POKA-YOKE

The goal of the error proofing in the context of *doing it right the first time* is to:

- Assess error proofing implementation at a supplier site.
- Identify error proofing opportunities during site assessments.
- Recommend corrective action during FMEA and 8D evaluations of detection rankings and minimizing or eliminating human-related problems respectively.

This means that the supplier has the responsibility to (a) correct any deficiencies that may be found during the site evaluation in such a way that they will *not* recur. In fact, the correction documentation must be available for a certain amount of time (this is defined by mutual agreement between customer and supplier), (b) to recommend corrective actions that will improve the detection ranking in the FMEA, and (c) correct deficiencies in the system where human behavior is depended upon, *e.g.*, inspection, orientation, and so on. To appreciate error proofing, one must understand the assumptions that are based on two essential attitudes about human behavior. They are:

- Mistakes/errors are inevitable.
- Defects can be eliminated or minimized.

An error is any deviation from a specified process. This means that all defects are created by errors. However, *not* all errors result in defects. The reason for this is that some errors are corrected in the process itself before they reach the next operation. An example of this is a 5-stand rolling mill. An error may be generated in the rolling process of the second rolling stand, but by the time it finishes the fifth rolling stand, the error has corrected itself, and there is no defect. Examples of errors are:

- The wrong option package is sequenced into the assembly.
- An oil sender unit is defective.
- A hose clamp is not positioned correctly during the assembly.
- A worn installation tool causes molding clips to be installed incorrectly.
- A door is left open on an assembly finishing line.

On the other hand, a defect is the result of any deviation from product specifications that may lead to customer dissatisfaction. To be a defect:

- The product must have deviated from specifications.
- The product does not meet customer expectations.
- The nonconforming product must have reached the customer.

Sometimes in this discussion (of error and defect), the notion of mistake comes into the conversation. A mistake is an error that occurs during processing. For our purposes, a mistake is viewed as an error. A comparison between an error and a defect is shown in Table 9.1.

So, in general terms, error proofing is a process improvement designed to prevent a specific defect from occurring. In some industries, the specific terminology of error proofing is reserved for design issues. Mistake proofing is reserved for process issues. The reader should notice that the Japanese word poka-yoke (ポカヨケ) does not make any distinction between the two. It simply means POKA – "Inadvertent mistake that anyone can make" and YOKE – "To prevent or proof" the process from any undesirable results. In literal terms, it means "fool proof." However, in the United States, "fool proof" is not acceptable because it is considered a disrespectful adjective for operators. Mistake proofing applies to:

- The application of tools and devices applied to a process to reduce the possibility of errors occurring.

TABLE 9.1
Comparison between Error and Defect

Error	Defect
The wrong option package was sequenced into assembly.	Vehicle equipped with wrong option package.
An oil sender unit is defective.	Engine oil/coolant leak.
A hose clamp was not positioned correctly during assembly.	
A worn installation caused molding dips to be installed incorrectly.	Squeak, rattles, or loose parts.
A door was left open on an assembly finishing line.	The door is damaged due to hitting an object.

- The application of tools and devices to a process to reduce the possibility of defects that have occurred continuing to the customer.
- The use of functional design features to reduce the possibility of parts being assembled incorrectly.

So, when one talks about error proofing, there is a distinction that has to be made. That is, are we talking about prevention or detection? See Figure 9.1. The way this question is answered will depend on the timing of the project (design or process), as well as whether a deficiency or nonconformance is identified in the course of conducting an FMEA or an 8D. Where does this assessment take place? Generally,

- *During normal site assessment*: When the assessment is conducted, the following observations should be made for identifying possible nonconformities:
  - All customer parts have documented design and process failure and effects analysis (DFMEA, PFMEA), and Control Plans (or Dynamic Control Plans – DCPs).

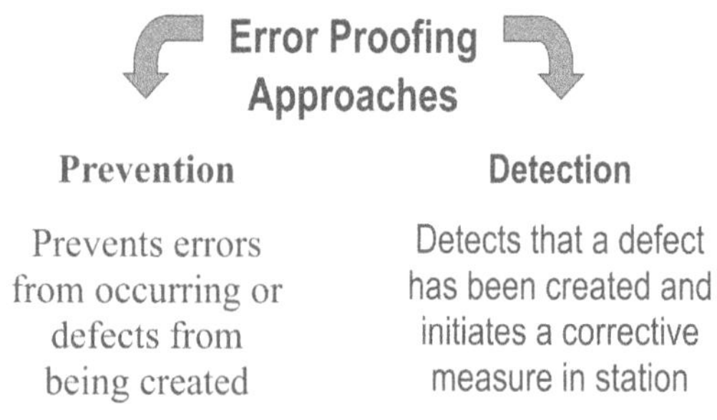

FIGURE 9.1 Comparison of prevention and detection approaches.

    - All defect detection areas are reviewed, and plans exist to move to defect prevention. There is evidence that defect prevention is replacing defect detection.
    - Supplier has a maintenance system that contains reactive, preventive, and predictive maintenance. The maintenance system supports process capability improvement as well as OEE expectations.
    - Supplier takes measures to minimize foreign material, chips, debris, contamination, excessive oil, *etc.*, where part quality can be negatively affected.
- *APQP reviews*: When this review is conducted, the following should be evaluated with extra attention:
    - DFMEA: Consider process improvement techniques (DOE, Robustness Methodology, Mistake Proofing (Poke-Yoke), 8D with high emphasis on Prevent Recurrence, *etc.*) to develop preventative and corrective actions for severity ratings of 9–10, severity ratings of 5–8, with an occurrence of 4 or more and high RPN items. [Note that the RPN is the least desirable criterion for evaluating FMEAs. The priority should be first on high severity, second on severity and occurrence, and third on the RPN. It is important to mention here that some customers evaluate FMEAs based on the value of the RPN. We do not recommend this approach.]
    - PFMEA: Address the impact of failures as applicable to each of the following items: Each part, subsequent operation, system, part, customer wants, government regulations, and operator safety.
    - Facilities/tools/gauges: Ensure that tooling and equipment design incorporates the recommended actions from the process FMEA.
- *FMEA reviews*: When reviewing the FMEA critical items for review are: Severity, potential cause, design/process controls, and their effectiveness.
- *8D reviews*: When a review for the 8D methodology is conducted, make sure that extra attention is on defining the root cause and escape point. Also, in step D5, check for verification – effectiveness; and D6 check for validation – date of implementation. Some items of concern should be the following. However, this is not an exhaustive list.
    - Critical features
    - Fit
    - Form
    - Function
    - Adherence to government regulations
    - Adherence to safety items
    - Customer satisfaction features
    - Customer concern issues
    - TGR/TGW (Things Gone Right/Things Gone Wrong) reports
    - Service bulletins
    - Field actions and stop shipments

So, based on what we have said so far, one can surmise that mistake roofing is a method/approach process improvement that is designed to prevent a specific defect from occurring. The goal is *zero defects*. This goal is based on three attributes. They are:

1. *Physical* – Install mistake proofing devices.
2. *Operational* – Ensure correct procedure, sequence, or execution.
3. *Philosophical* – Empowerment of workforce.

As a result of this method, benefits occur, and some of them may be the following:

- Enforces operational procedures or sequences.
- Signals or stops a process if an error occurs or a defect is created.
- Eliminates choices leading to incorrect actions.
- Prevents product damage.
- Prevents machine damage.
- Prevents personal injury.
- Reduces time to work on defective parts.
- Reduces cost of all materials used, including those used for rework.
- Reduces cost of time spent in rework.
- Reduces cost of time spent inspecting defective parts.
- Reduces lost production time spent on defective parts or rework.
- Increases customer satisfaction.
- Increases worker morale.
- And so on…

All these benefits are actually materialized because mistake proofing focuses on increasing value-added activities of products/processes as opposed to non-value-added activities. Value added is defined as any operation that changes, converts, or transforms material into the product that is sold to the customer, and the customer is willing to pay. Satisfaction is achieved! On the other hand, non-value-added activity is defined as any operation that does not transform in any way, shape, or form the product that is sold to the customer. Examples of non-value-added activities are:

- Inspection
- Storage/Inventory
- Downtime (including changeovers)
- Defects/Spoilage
- Waiting of any kind
- Rework of any kind
- Over/Under production
- Early/Late production

To take advantage of the poka-yoke methodology, improvement opportunities may come about when the following nine steps are followed.

1. Identify problem.
2. List possible errors.
3. Determine most likely error. (Verify that this is the error.)
4. Purpose multiple solutions.

5. Evaluate solutions' effectiveness, cost, and complexity.
6. Determine best solution. Verify that this solution will resolve the problem.
7. Develop implementation plan.
8. Analyze preliminary benefits.
9. Develop plan for long-term measure of benefits.

THE ZERO-DEFECT CONCEPT IN MISTAKE PROOFING

When one talks about zero defects, it is accepted as a goal of zero defects. Everyone knows that in the real world, there is no such thing as perfection in anything. The zero-defect concept is more often than not it is a definitional number. It is a quantitative statistic that everyone can support. A classic example is the parts per million (ppm) defect number. Of course, zero (0) is the perfect number. However, in most manufacturing facilities, both customer and supplier are happy if the ppm is around 50 ppm defects. In order for the zero-defect system to be effective, the following must be present:

- Source inspections.
- Mistake proofing devices.
- 100% inspection. (However, even 100% inspection is only 79% effective.)
- Immediate feedback and action.

So, how do we practice the zero-defect cycle? The answer is very simple. We have to follow a cycle similar to the improvement cycle of events. The reason for this is that "zero defects" is part of a continual improvement attitude/philosophy/culture in any organization. A typical comparison is shown in Table 9.2. Table 9.3 shows some of the specific questions that may be asked in five specific categories for demonstrating the zero-defect process.

TABLE 9.2
Comparison of Improvement Cycle, Zero-Defect Flow and Cycle

Improvement Cycle	Zero-Defect Cycle
• *Adherence:* The cycle begins with this stage. Its essence is to begin by following the current procedures, or the way things are done NOW. • *Analysis:* Analyze what is being done NOW and the results we are getting. • *Improvement:* Based upon the results of the analysis, we take improvement action. • *Standardization:* Standardize the successful improvements and continue the cycle.	• *Recognize defective work:* An item has been identified as a mistake. • *Conduct problem analysis:* The item is analyzed for a root cause and escape point. • *Improve:* Improvement is made based on a data-driven analysis. • *Stabilize:* Repeatability and consistency are assured. • *Adhere:* Assurance of doing exactly what was identified under stability. • *Zero defects:* Process is assured to be at zero defects as long as there is stability and as long as adherence is followed.

(Continued)

TABLE 9.2 (*Continued*)
Comparison of Improvement Cycle, Zero-Defect Flow and Cycle

The flow and rationale of the zero-defect cycle. It is imperative that when you are committed to zero-defects mentality, you must recognize that zero defects is a change in mindset. It is a philosophy that will take time for employee buy-in. Zero defects involves using a systematic approach to completing the task, inspecting the process, and achieving zero mistakes, zero defects, zero problems. In essence, zero defects means a mistake less environment with no exceptions. Absolutely no mistakes leave the work environment. The cascading process is:

Zero mistakes
  Zero defects
    Zero problems
      Zero flaws
        Zero errors
          Zero injuries
            Zero waste
              Zero down time

The result of this flow will result in

Stability → Continuous Flow → Synchronized Production → Pull System → Level Production

TABLE 9.3
Zero-Defect Process Overview

1	Defect	What is the product? What is the defect?
2	Mistake	Where was the defect found? Where was the defect made? What happens to the part in between? What is the most likely cause of the defect?
3	Root cause of mistake	Why? Why? Why? Why? Why?
4	Is short-term solution feasible?	If short-term action is necessary, then how can I stop the mistake from causing the defect at this point in time? If not, then the action is moved to the long term.
5	Long-term action	How can I ensure 100% inspection so this mistake will never occur again?

POKA-YOKE STEPS AND APPLICATION

As we already mentioned, poka-yoke, as foreseen by Mr. Shingo, was to improve quality. As such, the fundamental principles for mistake proofing that we all must adhere to are pretty much defined in eight steps. They are:

1. Build quality into the process.
2. Understand that inadvertent mistakes and defects can be eliminated.

3. Stop doing it wrong – start doing it right!
4. No excuses; think about how to do it right.
5. Anything over a 50% chance of success is good enough – Implement! The reason for anything over 50% is that statistically, one has a better than 50% chance of improvement from the status quo. Therefore, it is worth pursuing.
6. Mistakes and defects can be reduced to zero by working together and understanding both the task and process involved.
7. Build on synergy. Ten heads are better than one.
8. Seek out the true root cause and escape point of the defect.

These principles are very important; however, just like any system of improvement, with poka-yoke, there must be a standardized process of implementation to successfully improve the process. A general, generic approach is the 12-step approach. The steps are:

1. Identify and describe the defect.
2. Show the defect rate by charting the defect occurrence over time.
3. Identify where defect was found.
4. Identify where defect was made.
5. Describe the current process where the defect was made by detailing the standard procedures used in the operation.
6. Identify any errors or deviations from process standards where the defect was made.
7. Use the *5 Why or 3×5 Why* problem-solving technique to identify the root cause of the defect or error. If you can use the *3×5 Why* approach is even better.
8. Develop ideas for improving the process using *5-Why* or *3×5-Why* tools to eliminate or detect the error.
9. Improve the process by creating an error proofing device.
10. Measure/document results of error proofing.
11. Standardize the improvement.
12. Ask where else this improvement could be used?

Obviously, in some industries, these 12 steps may be modified to reflect their own needs. For example, Table 9.4 shows different steps for a typical automotive industry and a service organization, as well as the most common mistakes that poka-yoke may be used for best results.

Now that we know how to approach a poka-yoke opportunity, let us look at the optimum way that a process can improve. Table 9.5 shows a generic approach to selecting an appropriate level of control. Table 9.6 shows the levels that are the most efficient, and the right column shows some examples of production conditions where poka-yoke may be used effectively.

APPROACHES TO CORRECTING HUMAN MISTAKES

It has been said that "err is human." Indeed, humans do make mistakes, and that is why poka-yoke is a very powerful method to eliminate or at least minimize that "human error." Human mistakes, after all, (a) are part of someone performing a job or task (b) people will, on occasion, make mistakes, no matter how hard they

TABLE 9.4
A Typical Comparison of Poka-Yoke Approaches

Nine Typical Mistake Proofing Steps for an Automotive Organization	Seven Generic Steps for Service Organizations	Most Common Mistakes That May Be Improved with Poka-Yoke
1. Locate the defect and isolate the process that created it. 2. List all possible mistakes. 3. Determine the most likely mistakes. 4. Propose multiple solutions. 5. Evaluate each solution. 6. Choose the best solution. 7. Develop a plan for implementation. 8. Measure results/analyze benefits. 9. Update documentation affected by the mistake proofing implementation.	1. Identify and describe the defect. 2. Determine where the defect was discovered and made. 3. Detail the current standard procedure. 4. Identify deviations from standards. 5. Identify the Red Flag condition(s) where the defect occurs. 6. Identify the type of mistake- proofing device required to prevent the mistake or defect. 7. Create devices and test for effectiveness.	1. Omissions. 2. Processing mistakes. 3. Set-up mistakes. 4. Assembly/process mistakes (missing parts/information). 5. Wrong components (part, tool, *etc.*). 6. Preparation mistakes. 7. Procedural mistakes. 8. Adjustment mistakes. 9. Orientation issues. 10. Operator-depended processes.

TABLE 9.5
A Generic Approach of Selecting Appropriate Level of Control

	Level 1	Level 2	Level 3	Level 4	Level 5
Type	Autonomic operation	Automatic control	Self-regulation	Personal supervision, auditing, inspection	Information control
Use	Highly repetitive	Highly repetitive	Stable environment, established practices	Changing environment, cost of failure	Large area of responsibility
Example	Traffic light cycle	Temperature control via feedback loop - thermostat	Machinist, typist, assembly	100% inspection by the Q/C function	Warranty feedback

TABLE 9.6
An Effective System of Poka-Yoke That Works "Best"

Level	Process	Approach	Production Condition
Level 1	People	Job is standardized and stabilized.	• adjustments
Level 2	System	Many defects are produced but detected.	• tooling/tool changes
Level 3	Subsequent station	Several defects produced but detected.	• lots of parts/mixed parts
Level 4	At the station (where work is done)	Possibly one defect produced and stopped at station.	• multiple steps
Level 5	Process parameters (manufacturing)	Mistake corrected, defect not produced.	• infrequent production
Level 6	Design (product/ service)	Error eliminated.	• lack of effective standards
			• symmetry - parts that are mirror images
			• asymmetry - parts that look alike but are not
			• rapid repetition
			• high production volumes
			• environmental conditions

try - they can forget (c) mistakes can lead to defects, and (d) people must be supported by system that says mistakes can be prevented. However, when the focus is on fixing the person (the *WHO*), we are "blaming someone" instead of fixing the mistake. The focus should be on identifying the facts and eliminating the mistake through the appropriate level, as shown in Table 9.6.

To approach the defects, we must first track and evaluate their significance to the system. Typical traditional tracking is based on the flow of: mistake → mistake not corrected → defect. Specifically, typical mistakes are identified by:

- Quantity
- Kind
- Percent
- Scrap rate
- Rejects per defect

With mistake proofing, the focus of tracking must be by: (a) The point at which the defect is discovered, and (b) the point at which the defect occurred (escape point). The flow is quite different and is based on source inspection. That is:

mistake → mistake discovered → feedback and action → zero defects

So, how can we differentiate the traditional approach from the mistake-proofing approach? In the first, we depend on self-check, and in the second, we depend on source inspection. Self-check means that the operator performing work checks on an item before passing it to the next process. This is self-inspection by the operator through sensory and visual cues. On the other hand, source inspection is taking preventive action, at the mistake stage, to prevent mistakes from turning into defects, rather than taking corrective action after defects have occurred. It is a preventive approach as opposed to a detection approach.

The focus, therefore, for any improvement using poka-yoke methodology is to develop sound approaches for controlling:

- *Value Added*: Value-added activities are those which add form or function to the product. Any activity that does not add form or function is considered non-value-added waste.
- *Just-in-time information*: The right information, in an appropriate form, in the hands of the people who can act on it, precisely when it is needed.
- *Source inspection*: (a) How to discover mistakes in conditions that create defects, and (b) how to prevent mistakes from turning into defects.
- *The 3 actuals*: To be effective, process control, inspections, and information sharing must be understood in terms of the 3 actuals: (a) The actual *place* or location an operation occurs (b) the actual *people* in that place, and (c) the actual *process* occurring in that place.
- *A visual language*: It is a way of communicating fast and accurate information by: (a) Recognizing problems and abnormalities, (b) taking quick corrective action, (c) ensuring necessary learning, and (d) preventing problems from recurring.

In order for the strategies to be effective, the first requirement is that there must be an environment/culture in the organization that does not assign blame to any one individual but pursues issues in design and or process. A no-blame environment means that (a) problems are the fault of the process, (b) work occurs in a dynamic world, (c) communication is vital to success, and (d) everyone is responsible for *no-blame*. Focus the questions on *why* did it happen? or *what* happened? Or *how did* it happen? Avoid asking *who* did it?

The second requirement is to cultivate the zero-defect mindset throughout the organization. In order to maximize the results of using zero-defect system for achieving zero defects, the following steps must be included in the process: (a) Reduce the amount of time between the discovery of defects and determining where the error occurred, (b) use 100% successive checks, and (c) use 100% self-checks.

Poka-yoke, as we have already mentioned, is a way to minimize or eliminate defects. However, there are two types of defects:

1. Isolated – They happen once.
2. Serial – They happen more than once.

Both types can be either: (a) Processing defects – process dependent, and (b) material defect – material dependent. In both conditions, they are created by either human conditions or production conditions. To identify them, more often than not, we depend on inspection, even though we know there are many shortcomings. For example:

- Although it is necessary to have efficient inspection operations, they are of little value in the process. Even the most efficient inspection operations are merely efficient forms of waste.

- Inspection involves the examination of a product or process in order to detect mistakes, defects, or problems.
- Inspection only makes sense from the company's standpoint, not the customer's.
- The customer does not care if the item is inspected. The only concern of the customer is that the item be flawless.
- To the customer, inspection is a *non-value-added.*

So, even though inspection comes with some concerns, we still depend on it, for they do provide significant information based on *some* judgment. However, depending on the situation, there are three types of inspection.

1. *Information inspections*: They are based on observation to: (a) Generate feedback regarding quality issues so that corrective actions can be taken, (b) help reduce defects, but do not prevent them, (c) focus is still reactive by correcting the problems after they occur, and (d) information inspections take two forms (i) pass or (ii) fail.
2. *Judgment inspections*: They occur after the mistake and or the defect has been made. They are reactive in nature. In essence, a source inspection is: (a) Based on knowledge of the standards and a comparison to those standards, (b) the inspections are reactive by determining the problems after the fact, and (c) it is difficult to obtain consistency because the inspections are based on subjective human judgment rather than objective measurements.
3. *Source inspections*: These are done at the source of the task. Therefore, the inspection is capable of detecting mistakes as they occur and eliminating them.

All three types of inspection are a form of a check. This check can be a self-check or a successive check. Regardless of what type of check is conducted, the intent is to discover defects or reducing future defects. Specifically, the inspection to discover defects is concerned with (a) products compared with standards, and defective items are removed, and (b) sampling inspection when 100% inspection is too much trouble. The inspection to reduce future defects is based on statistical quality control activities such as SPC – control charts – to anticipate and avoid defects based on the assumptions of the notion that some level of defects is inevitable (Type one error) and the interest is in results rather than the cause. A comparison of these two approaches is shown in Table 9.7. So, is there a clear way to identify mistakes and defects? Yes, it is. The distinction will be easier if the following information is available.

- *Product*: What is the product (and or service)?
- *Defect*: What is the defect?
- *Mistake*: What is the most likely cause?
- *Detection location*: Where was the defect discovered?

TABLE 9.7
A Comparison of Self and Successive Checks

	Self-Checks	Successive-Checks
Description	The inspections are performed by the individual who performs the task.	The inspections are performed by the individuals who performed the task.
Advantages	• Represent 100% inspection. • Feedback is immediate and, therefore, corrective.	• May detect mistakes overlooked during the previous operation(s). • May detect errors missed during the self-check. • Encourage teamwork.
Disadvantages	• Conscious and or unconscious compromises may be made on quality. • The inspections may not occur due to forgetfulness, time, or motivation.	• Feedback may be slow. • Corrective action occurs after the mistake is made. • May foster blame.

- *Source location:* Where was the mistake made?
- *Duration between source location and detection*: How long between mistake and defect? (time, work stations, processes, and so on).

A comparison of mistakes and defects is shown in Table 9.8.

In addition to inspection practices to improve quality in any poka-yoke method, one will also find the use of mistake proofing devices. Mistake-proofing devices are devices made up of elements when followed, cause a predictable and desirable result and, when not followed, result in a predictable defect or waste. Mistake proofing

TABLE 9.8
A Comparison between Defect and Mistake

Defect	Mistake
Quartz watch does not work	Dead Battery
No hole drilled	Broken drill
Undersized hole	Worn drill bit
Inaccurate expense report	Typographical error
Computer application glitch	Software error
Car won't start	Transmission not in "park"
Peeling paint	Improper paint preparation

devices should be developed by a team, developed for a specific application, be simple, reliable, and inexpensive. They are used to:

- Prevent mistakes from occurring.
- Detect mistakes that have occurred and prevent defects.
- Detect defects that have been made before the devices are installed.
- Able to perform 100% inspection.
- Designed to detect only one mistake.
- Able to provide prompt feedback and action.

There are many types of mistake proofing devices for many applications. They may be very technical and complex but also very simple. Typical devices are:

- Guide/reference/interference rod or pin
- Template
- Limit switch/microswitch
- Counter
- Odd-part-out method
- Sequence restriction
- Standardize and solve
- Critical condition indicator
- Detect delivery chute
- Stopper/gate
- Sensor
- Mistake proof your Error Proof device. This is the most misunderstood item in device implementation. This addresses the lack of good maintenance that can render the most brilliant error proofing device useless or even dangerous. Dangerous, because if precautions are not taken, you can assume that the device is watching your quality or safety, when it is not. To avoid this, you must regularly maintain every non-mechanical error proofing device. This includes sensors, limit switches, counters, gates and stoppers, and any other device that relies on electricity, temperature and pressure gauging, or tolerances. Your device is reliable only to the extent that these pre-determined specifications are precisely maintained.
    - Eliminate the condition
    - Redesign for symmetry
    - Redesign for asymmetry
    - Sticky notes on components
    - Electronic eyes
    - Components that will only attach one way
    - Limit switches
    - Probes
    - Proximity switches
    - Profile plates
    - Alignment tabs
    - Broken tooling indicator

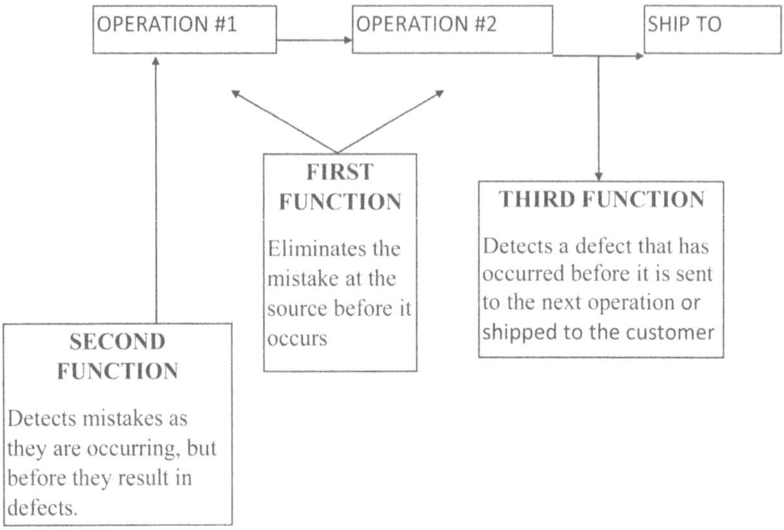

FIGURE 9.2 Functions of mistake proofing devices.

- Color striping
- Photo cell eyes

The typical functions of mistake proofing devices are shown in Figure 9.2. Some of the devices are the result of regulatory functions, but all devices are of three types. They are:

1. *Warning method*: Signals the need for feedback and action when abnormalities occur (least effective). When an abnormality occurs, feedback is provided by signals or visual controls, *e.g.*
   a. Lights
   b. Alarms
   Once these devices are activated, the mistakes and or defects will continue to be produced until an operator responds.
2. *Shut down method:* Shuts down or halts the operation when abnormalities occur. When an abnormality occurs, the operation automatically shuts down, and before it starts up again, it needs the intervention of an operator
3. *Control method*: Automatically corrects the mistake condition (most effective). When an abnormality occurs, the mistake proofing device automatically corrects the mistakes

RED FLAG CONDITIONS

A red flag condition is a condition in the manufacturing/service process that commonly provokes mistakes. Typical red flag conditions are:

- Adjustments
- Symmetry

- Tooling/tooling changes
- Asymmetry
- Dimensions/specifications
- Critical conditions
- Rapid repetition
- High/extremely high volume
- Many/mixed parts
- Multiple steps
- Infrequent production
- Lack of an effective standard
- Environmental conditions:
  - Material/process handling
  - Foreign matter
  - Poor lighting
  - Housekeeping

How do you identify red flag conditions? There are many ways depending on the process. However, the following steps may help in identifying red flag conditions.

- Identify and describe defect.
- Determine where the defect was discovered or made.
- Detail the current standard or procedure.
- Identify deviations from standard.
- Identify red flag conditions where defect occurs.
- Identify the type of mistake proof device required.
- Create the device and test it for effectiveness.

CORRECTIVE ACTION

Now that the defect has been identified, there must be an action developed to eliminate it. After all, taking corrective action means not using the same old solution, or the solution is the result or accomplishment that will be better than what you had to start with. There are two types of corrective action. They are:

1. *Short-term corrective action:* They aid in keeping things running, but may not keep the problem or defect from recurring.
2. *Long-term corrective action:* They prevent the defect or problem from recurring.

To accomplish either one of the options, a strong rationale and mindset in the organization must be prevalent. This means that the organization must exercise the following on a daily basis and be on the lookout for any deviations:

1. Prevent defects from passing on to future products.
2. Detect mistakes and give feedback.
3. Correct or eliminate mistakes.

One of the *best* ways to maximize the corrective action is by having appropriate and applicable support. That appropriate and applicable support is manifested through the team. Specifically, support means:

- Having appropriate and applicable teams.
- Having appropriate and applicable training.
- Having a standardized methodology.
- Having the commitment and support of management.

Appropriate and applicable team means that all participants are working together for a common *fix* of a specific problem. Specifically, the team must be cohesive in the following areas:

- Treat each other with respect.
- Stay focused on the task.
- Offer suggestions, do not be afraid to speak up.
- Constructively criticize ideas, do not criticize people.
- Listen to others until they are finished.

This cohesion is very important because more often than not, the composition of the team is very diverse in experience, education, and title of position. A typical composition may be made up of:

- Support personnel
- Operators
- Skilled trades
- Supervisors
- Engineers

The composition of the appropriate team members is of profound importance because the first decision must be made whether or not the resolution of the problem at hand has to be with internal or external activities. Internal activities must be performed while the machine is shut down. On the other hand, external activities can be performed while the machine is running. Obviously, the best option is to do work while the process is operating. One way to analyze the situation is by brainstorming and specifically:

- Analyzing each element.
- Asking the 5 whys or the modified 3×5 why's.
- Resolving if the element can be made external.

If brainstorming is used, then:

- Record all ideas.
- Do not emphasize ONLY hi-tech solutions.
- Inexpensive solutions are often viable.

- ALWAYS maintain a questioning attitude.
- Zero changeover time is possible.

Do not forget to mention the issues of scheduling. For example:

- Poor schedules from the scheduling activity.
- Production ignores schedules from the scheduling activity.
- Equipment breakdowns.
- Material and or personnel shortages.
- Quality issues.
- Product complexity.

Once the team has evaluated the area of concern and disposition of a corrective action is about to be identified and recommended it is imperative that the action taken *must* be based on the principles of employee empowerment (*i.e.*, team members must have been given authority and responsibility of their situations) and the final decision *must* be consensus driven. Empowerment is very critical for it allows the mistake proofing team to:

- Make changes to their work environment.
- Take ownership of the product they manufacture.
- Support continual improvement in the workplace.

Consensus, on the other hand, is reached through a diligent discussion without fear of ridicule and or intimidation by the other team members and or management. Specifically, consensus is reached through the process of:

- Discussing thoroughly each idea.
- Eliminating duplicate ideas.
- Work as a group to pick the best idea.
    - Do not let one person dictate the outcome.
    - All team members should agree on the name.
- Decision is reached by agreement, NOT voting or majority for the best idea.

PRESENTING FINDINGS

When the team finishes studying a problem and has indeed figured out a solution to the specific defect, now is the time to present the results to the management team who is, in fact, the final judge as to whether or not the solution is acceptable. This means that the mistake proofing team must do some preparations. The obvious one is to start by:

- Deciding what information, you are going to present.
- Deciding how the information will be presented.
- Deciding who will present (may be more than one person).
- Is there a need to practice the presentation?

Within these four categories, the predominant factors are the consequences of the *defect* (the reason for the study) and the *specific fix* with the associated potential benefits. The first one is addressed through – at least the minimum information – by focusing on:

- Mistake proofing location.
- Potential defect (escape and detecting).
- Mistake that could cause this defect.
- How is this mistake prevented?
- Appropriate recommendation (prevention, detection, or correction).

The second one is addressed by summarizing the benefits with quantifiable data. Without data, the mistake proofing work will not be accepted. The benefits of any mistake proofing undertaking are generally traditional costs and savings based on the number of defects eliminated. Do not include unquantifiable costs and undesirable situations as part of the presentation – just because they sound good. They will distract from the focus and waste time. Typical unquantifiable costs and undesirable situations are:

- *Business costs:* Cost of additional energy and equipment required to manage a poor-quality production environment.
- *Liability costs:* Cost of losing business due to poor product quality or inefficient compliance with government regulations and safety.
- *Cultural costs:* Reduction of worker morale and work standards. In some cases, community acceptance may play a role in the decision.

NEXT STEPS

The knowledge and skills you gained from conducting the study should be used (a) as a regular part of your work, and (b) diffused throughout the organization as applicable. Remember that success begets success! Remember also that mistake proofing should come from everyone. As a consequence, mistake-proofing applications should be *simple*. Recommended next steps are:

- Follow through on mistake proofing projects.
- Use a mistake proofing opportunity and tracking form to develop other mistake proofing opportunities. A typical format for this is to:
    - I. Identify the problem.
    - II. List possible errors.
    - III. Determine the most likely error.
    - IV. Propose multiple solutions.
    - V. Evaluate solutions' effectiveness, cost, and complexity.
    - VI. Determine the best solution.
    - VII. Develop implementation plan.
    - VIII. Analyze preliminary benefits.
    - IX. Develop plan for long-term measure of benefits.

Typical tracking items may be in the form of:
- Percent of errors proofed.
- Percent of errors detected vs. percent of errors prevented.
- Percent of errors attributed to design.
- Percent of errors attributed to process.

- Bring your knowledge and skills of error-proofing back to your work group.

AUDITS

OVERVIEW

The general definition of an *audit* is a planned and documented activity performed by qualified personnel to determine by investigation, examination, or evaluation of objective evidence, the adequacy and compliance with established procedures, or applicable documents, and the effectiveness of implementation (Stamatis, 2002). The term may refer to audits in accounting, internal controls, quality management, project management, water management, safety, energy conservation, and so on.

On the other hand, *auditing* is defined as a systematic and independent examination of data, statements, records, operations, and performances (financial or otherwise) of an enterprise for a stated purpose. In any auditing, the auditor (the person who does the audit) perceives and recognizes the propositions before him for examination, collects evidence, evaluates the same, and on this basis formulates his judgment which is communicated through his audit report. The purpose is then to give an opinion on the adequacy of controls (financial and otherwise) within an environment they audit, to evaluate and improve the effectiveness of risk management, control, and governance processes. Generally, it is a system oriented.

Specifically, *quality audits* are performed to verify conformance to standards through review of objective evidence. A system of quality audits may verify the effectiveness of a quality management system. This is part of certifications such as ISO 9001. Quality audits are essential to verify the existence of objective evidence showing conformance to required processes, to assess how successfully processes have been implemented, and to judge the effectiveness of achieving any defined target levels. Quality audits are also necessary to provide evidence concerning the reduction and elimination of problem areas, and they are a hands-on management tool for achieving continual improvement in an organization (Stamatis 2004).

To benefit the organization, quality auditing should not only report nonconformance and corrective actions but also highlight areas of good practice and provide evidence of conformance. In this way, other departments may share information and amend their working practices as a result, also enhancing continual improvement (Stamatis 2012).

CS can benefit from audits because they deal with immediate problems and or positive situations. The opportunity for fixing and or improving the CS is very real, and the organization will miss that opportunity for improvement. To put it in simple language, the audit for any CS situation is an easy and efficient way of gaining positive result(s) from low-hanging fruit. For a very lengthy discussion on audits, see Stamatis (2021, 2021a)

TYPES OF AUDITS

There are many types of audits for different industries. However, all audits are divided into three categories. They are:

1. *First party audit.* An organization audits itself. The objectives usually are to:
   - Validate a process. The process may be an activity that adds value to the product or service. Identify any gaps in what the process must do as opposed to what is being done.
   - Validate project implementation activities based on past deficiencies or current action plans for nonconformities.
   - Validate product characteristics or performance requirements.
   - Validate that nonconformities have been addressed.
   - Validate training, equipment capabilities, process settings.
2. *Second party audit.* A customer audits the supplier. The objectives are to:
   - Validate supplier organization processes used to provide a product or service.
   - Validate supplier product or service characteristics or performance requirements.
   - Validate supplier process capabilities, including capacity.
   - validate conformance to contract requirements.
   - Validate material sources and traceability.
   - Validate that defects and nonconformities have been addressed.
   - Validate that a Quality Operating System (QOS) exists, is reviewed, and implemented.
   - Validate that an internal audit is conducted and appropriate corrective actions are taken for any deficiencies.
3. *Third party audit.* An independent certification body conducts the audit. This type of audit is certified usually for three years. At the end of this time frame, it needs to be conducted again. Some certified bodies require an external audit to be conducted once a year. The objectives are to:
   - Approve/disapprove process for certification.
   - Approve/disapprove product or service for certification.

In all three categories of auditing, it is imperative that the scope of the audit is identified up front. The second imperative of all audits is the need to understand the process and product/service the auditor is going to audit. Reviewing the procedure, specifications, and records is a good starting point. If there is no procedure, you may need to ask the auditee to provide a description of the process or processes (a process flow chart may help). Ask questions, talk to people to get the information you need.

There are several tools available to help you to understand the process. They include:

- Process flow diagrams, flowcharts, or process mapping
- Cause-and-effect diagrams, turtle diagrams, 3×5 Whys
- Tree or functional diagrams
- FMEA (DFMEA and PFMEA) results
- Training documents (employee matrix)
- Inspection checklists
- Procedures and instructions
- Bill of materials, quantities, specifications
- Relevant standards

In addition to the tools, the auditor must not underestimate the value of understanding the process input and output criteria for good performance. Appropriate criteria may include:

- Specifications, lists
- Drawings, pictures, diagrams
- Planned arrangements for process approval
- Approved equipment and qualification/certification of personnel
- Test procedures
- Inspection methods (sampling, frequency)
- First article inspections
- Contract or regulatory requirements

In any audit, regardless of type, the history of the process plays a major role in identifying trends and gaps in corrective actions. Therefore, the auditor must review as part of their planning at least the following:

- Nonconformance reports and trend analysis
- Internal and field failures (how they happen and what was done)
- Corrective actions (check for appropriateness and timeliness)
- Process/product changes, date and nature of changes (how change was handled?)
- Operator/technician changes (are these changes completed with appropriate personnel?)
- Revalidation history (review history for continual improvement in the organization)
- Customer complaints (how are they handle? Is there a standardized method of problem-solving methodology? Are they completed in a timely fashion?

PERFORMING AN AUDIT

All audits have a base of comparison. Usually, they start with a standard and then add the specific requirements of the customer, as well as the organizational standards included in procedures and instructions of the organization itself. Details of how to conduct audits may be found in Arter (2003), Arter *et al* (2012) and Stamatis (1996). Generally, auditors collect information by using open-ended questions to gather data about process inputs, outputs, and the process elements (people, environment, equipment, material, measuring and method). They report not only weaknesses but also strengths in the organization based on objective evidence. Objective evidence are considered to be written documentation, personal interviews with more than one person, documented observation with a witness, or a combination of these. An example of a typical preparation for an audit may be found in Table 9.15. The example is not an exhaustive audit, but hopefully the reader will see the flow of some requirements, the criteria for the requirements, and the items that the auditor should look for.

If the audit is conducted to validate conformance to a standard such as the ISO/TS 16949, then the auditor compares what he finds in the organization against the elements or clauses of the standard. This same strategy is used in product/service audits to determine if the product conforms to specified requirements that are in a clause or element of a standard, specification, or condition document (this is very important if the customer has customer-specific requirements). A product audit can include validation of product characteristics as well as performance requirements.

Verification audits are generally conducted to validate error-proofing actions for specific processes and are generally conducted in reverse order, since the verifiable links are more obvious in the corrected process. In the auditing world, when the focus is on quality management audits, there are four approaches to identify gaps, correct (close the gaps) and monitor the improvements. They are:

1. Product audit
2. Process audit
3. Layered audit
4. Error proofing

Product Audit

Product audit is a planned and systematic study of finished products prior to delivery, which must be regarded as a normal part of the quality assurance activities. In some cases, the customer may initiate such a product audit to be carried out jointly with the supplier, such as in situations of a start-up of just-in-time deliveries.

The audit is to be performed with the help of written instructions and check lists, which has been drawn up by the supplier. Instructions shall be based upon the technical specification and other valid specified requirements.

A product audit has to be performed from the customer's perspective or as commonly known as *with the eyes from the customer.* Measures must be duly taken whenever deviations have been discovered or when a negative trend is observed. Alarm and action limits must be set.

Sometimes, a product audit is called a *source inspection audit*. The differences are very minor, if any, and most professionals in the quality field use them interchangeably. In any case, for those who see some differences, here is an overview.

Source inspection audit is the inspection of purchased products and process records at the supplier's premises to verify compliance with requirements before shipment to the client. Examines the customer-purchased product that is performed at the supplier's facility to verify product integrity and conformance to specified requirements prior to delivery. It involves the process of examination of a finished product before shipment in order to test the conformance of sub-assemblies may be obscured and thus impossible to subsequently inspect as the assembly is built up. Also, serves as the first inspection of a production piece that checks for compliance with engineering drawings. Typical applications are tests, castings, forgings, and machine parts to prove tooling before embarking on production runs, and electronic assemblies in order to prove the effectiveness of the production process before committing to order quantities. Pre-cap inspection typically takes place prior to sealing or encapsulating high-reliability microelectronic components in cases where compliance with military space or international specifications is critical. A source inspection audit should be used when:

- Companies and organizations see the need to examine a customer-purchased product at the supplier's facility with the intention to verify product integrity and conformance to specified requirements prior to delivery.
- Companies and organizations need to inspect a finished product immediately before shipment.
- Companies and organizations need to examine and ensure that the conformance of subassemblies may be obscured and impossible to inspect, and subsequently are thoroughly examined.
- Organizations need to inspect the first production piece to check for compliance with engineering drawings.
- Organizations intend to inspect castings, forgings, and machined parts in order to prove tooling before embarking on production runs.

Some of the benefits of Source Inspection Audit are:

- Enables the reviewing of the supplier's processing records to establish that all plant manufacturing, inspection, and test operations have been carried out and properly recorded.
- Enables the checking for evidence that any measuring equipment used in prior inspections or tests was under a system of calibration control, and within calibration at the time the inspections or tests were performed.
- Ensures the verification of the product complies with pertinent engineering drawings.
- Ensures that product identification marking or labeling requirements set out in drawings or associated specifications are properly observed.

The reader should be very cognizant of the fact that a product or source inspection audit is not very dependable and should be avoided whenever possible. The integrity of the product quality should be measured with statistical approaches, including SPC. In any case, both of these approaches do have a business impact, which:

- Determines the status of any related nonconformance report, waivers, or deviations, and how these bear upon the acceptability of the product.

- Ensures that products inspected in the process conform to assembly drawings to the appropriate degree for the stage of manufacturing at which they are inspected.

Process Audit

Process audit is a planned and systematic examination of the equipment, procedures, and activities according to the documented operations with respect to capability, application, and effectiveness. As a preparation, an audit plan shall be drawn up incorporating the organization of the audit, the participants, the extent, and the basic information. The extent of the audit can, for example, be identified by:

- Production area
- Department
- Workshop (floor) area
- Production line
- Special process

On the other hand, the basis for the audit can consist of:

- Product specifications
- Process descriptions
- Written procedures
- Questionnaire/check lists
- Standards (international and or specific customer requirements)

A process audit is generally used by customers to evaluate suppliers. However, the strength of the process audit is in the fact that organizations can use this type of audit to find nonconformances within their processes and correct them.

Arter (2003) has identified seven steps in the process approach to auditing. They are:

1. Define the products.
2. Define the processes used to make those products.
3. Analyze and understand those processes.
4. Develop objective evidence needs to explore the processes.
5. Gather field data during the actual audit.
6. Analyze the gathered data by sorting into piles.
7. Present the information to the process owners.

In order to use the process approach, the auditor must first understand how the business processes relate to the objectives of the organization. Then, data are gathered to see how these processes are being controlled. Finally, conclusions must show how the identified strengths and weaknesses affect the operation of the business. A typical partial process audit form is shown in Table 9.9.

TABLE 9.9
Partial Process Audit

Process Audit										
Department Audited:									Date:	
Auditor's Name:									Shift:	
Note: All questions are to be answered with a "Y" or "N" for yes and no respectively under the corresponding day's box. ALL nonconformances *must* have corrective action recorded under the specified column in this form with the date (MM/DD/YY)										
Item no.	These items must be checked in every audit	Audit area	M	T	W	T	F	S	S	Corrective action
Section 1: Operator Instructions										
1	Are operator instructions accessible at the work stations and made available without disruption to the work being performed?									
2	Do operators understand their work instructions and is the work being performed follow the work instructions step by step?									
3	Do all operators performing containment activities have work instructions and have they been trained to the work instructions?									
4	Do all operators have a clear understanding of their jobs? Do they know the right thing to do, are they doing them correctly, and can they demonstrate?									
5	Are operators filling out required forms and documentation and turning them into their supervisors per their work instructions?									
Section 2: Product identification, Housekeeping and Workplace Organization										
1	Is all product in the work area properly identified with product status clearly shown per work instructions and procedures (Identify the actual #)?									
2	Are all containment, quarantine and product staging areas physically identified?									
3	Are all work areas maintained in a state of order, well organized with good housekeeping?									
4	Is suspect product properly identified per work instructions and procedures (identify the actual #)?									
And so on										

In Table 9.9, the reader will notice that the process audit depends on questions to make sure that the procedures and processes that are agreed on during a selection audit process are followed. The complete document must identify nonconformances in the manufacturing process, engineering change process, invoicing process, quality process, and the supplier/shipment process. These audits are analyses that (a) identify internal nonconformances to standards, procedures, instructions, and (b) are done to document the relationship between different companies in order to verify compliance of a supplier's products and processes.

A process audit should be conducted when:

- Companies need a supplier audit to verify their compliance with their supplier's products or processes.
- Companies need to improve their Return on Auditing (ROA).
- Companies would want to manage their supply quality management system so that they can know that their suppliers are in compliance.
- Companies struggle with reallocating their resources such that they cannot keep current projects on schedule while their engineers are out in the field conducting supplier audits.
- Businesses need to conduct mutual and beneficial audits of their company and suppliers.
- Companies or businesses want to create an integrated auditing process aligned with internal goals and supplier commitments.
- Companies require flexibility and consistently structured supplier audit services so that they can execute a value-adding audit program.
- Companies want to improve their quality level.

Some of the benefits of this type of audit are:

- Recognition of implementing an overall quality system.
- Recognition of establishing documentation control.
- Recognition of the need to control business operations and records.
- Recognition of establishing a measuring system for customer satisfaction.
- Implementation of a business management review for key processes and indicators.
- Implementation of communication policies to correct nonconforming products.
- Implementing corrective actions in a timely manner.

Just like everything else that is done in any organization, so does the process audits have business impacts. Some of them are:

- Information and product improvements can be shared for common gain among departments, divisions, and different geographical facilities.
- Accomplishments of organizational measurement that can take place, risk mitigation exercises, both proactive and reactive.

- When there is an existing stand-alone supplier quality management function, strong relationships with suppliers can be created from a customer's perspective.
- Knowledge can be integrated with deliverance of profit-generating opportunities for both organizations and the exploration of additional, above and beyond, contrasting business opportunities. This can be achieved through the creation of a supplier quality management function for all those involved, such as account managers, supply chain consultants, and supplier performance managers. [This is what supplier development is all about!]

Layered Audit

Layered process auditing, also known as LPA, is an increasingly popular quality tool developed for manufacturing management – especially in the automotive industry. When utilized properly, LPA will drive cultural change throughout an organization to improve quality, reduce scrap and rework, and reduce customer rejections. But what exactly are LPA?

Layered audits represent a formalized system focusing time and energy toward ensuring high-risk processes and error-proofing devices do not exhibit a lot of variation and are working properly. A LPA system is comprised of three critical elements:

1. A set of audits focused on high-risk processes.
   An example of an audit form for high-risk processes is shown in Table 9.10.
2. Layers of auditors from all areas of management who perform audits.
   A typical format for this type of layered audit is shown in Table 9.11.
3. A system of reporting and follow-up to ensure containment as needed and drive Improvement.
   A typical form of a layered audit is shown in Table 9.12, indicating the documentation and progress from shift to quarterly improvement.

AIAG's document CQI-8 provides the details of a generic layered audit. Fundamentally, what the CQI-8 describes is that the audit may be scheduled on a daily, weekly, monthly, or quarterly to make sure that the critical to quality (CTQ) characteristics are met from the customer's perspective. No matter what the frequency, each audit should last anywhere from 10 to 30 minutes, and they should have the appropriate level of management. Of course, each level of management will be investigating/evaluating the items that are within that layer's span of control. The higher the layer, the less tactical and more strategic the items reviewed.

It is imperative to understand that all layered audits must focus on areas in processes where deviation (variation) may cause a product to be a high risk for either the customer or the producing organization or both.

TABLE 9.10
A Typical Format for a Layer Audit Focusing on High-Risk Processes

Part Name:							
Part Number:			Shift:				
Customer:			Week Ending:				
Department:			Other:				
Operation 10	Broach	Mon	Tue	Wed	Thu	Fr	Sat
	Is runout within specs?						
	Has perpendicularity been gauged? Spec is 20 lines.						
	Is there any broach tear? Check one part per broach.						
Operation 20	Olofssons						
	Is the ID on size?						
	Is the ID runout within specifications?						
	Is the ID taper within specifications?						
Operation 30	Hones						
	Has the ID-micro been checked and the micro log completed?						
Entire Department	Are operator's verification sheets completed daily by all shifts?						
	Are the tagging procedures being followed?						
	Are there any temporary alert notices posted that have expired?						
Operation A	Assembly						
	Are the builders checking rotation and marking parts? (touch point)						
	Have the eight-hole sprags inspected for presence of 8 holes and identified with an orange marker?						
	Are bar coding labels attached to shipping container in correct location?						
	Are shipping containers checked for correct count?						
	Machining operations total NON conforming items						
	Assembly total NON conforming items						
List any quality or manufacturing concerns in any operation:							

TABLE 9.11
A Typical Form of a Management Layered Audit

A Typical Layered Format

Level	Assigned Management Category	Assigned Management Personnel	Audit Assignment	Process A	*Etc.*
First layer of management	Supervisors	Process A Mng	Own Department	1 per day	1 per day
		Process B Mng	Own Department	1 per day	1 per day
		Assembly Supr. Mng	Own Department	1 per day	1 per day
		Quality Sup. Mng	Own Department	1 per day	1 per day
Second layer of management	Plant staff and Plant Mng	Quality Mng	Rotate Departments		
		Maintenance Mng	Rotate Departments		
		HR Mng	Rotate Departments		
		6 sigma/Lean Training coordinator	Rotate Departments		
		Assistant Plant Mng	Rotate Departments		
		Assembly Mng	Rotate Departments		
		Process A Mng	Rotate Departments		
		Process B Mng	Rotate Departments		
		Shipping Mng	Rotate Departments		
Third Layer of management	Corporate Management	VP of Quality	Rotate Departments		
		VP of Operations	Rotate Departments		
		Systems Engineer	Rotate Departments		

NOTE: (1) If there is no supervisor, then the area manager shall be considered the first layer
(2) Of course, other management representatives may participate as appropriate

TABLE 9.12
A Typical Layered Audit Indicating the Documentation and Progress from Shift to Quarterly Improvement

Item	Shift		Daily		Weekly			Monthly			Quarterly		
Part/Product													
	Op	Insp	TLead	Sup	P Mng	QA Mng	OP Mng	EX. Mng	Cor Mng	P Mng (optional)	Pres	P Mng	QA Mng
Part/Product													
Error proofing verification													
First piece inspection													
Last piece inspection													
Standard work instruction present													
Operator training tracking sheet													
Safety issues													
Process													
Set up													
SPC compliance													
Tooling approval (Sign-off)													
Quality gate data													
System													
Preventive Maintenance													
Calibration													
Lot traceability													
Housekeeping													
Voice of Customer													
6-panel posted													
Action plans updated and on time													
Customer delivery performance													
Other relative info													

Legend: Op = Operator; Insp = QA Inspector; TLead = Team leader; Sup = Supervisor; P Mng = Plant Manager; QA Mng = Quality Manager; OP Mng = Operation Manager; EX. Mng = Executive Manager; Cor Mng = Corporate Manager; Pres =President

An LPA should be conducted when the organization wants to ensure high-risk processes and error-proofing devices do not exhibit a lot of variation and are working properly. Typical areas are:

- Assembly area
- Manufacturing operations
- Shipping/Receiving
- Repair/Rework area
- All operations and other support functions

Some of the benefits of using an LPA are as follows:

- Minimizes variation in processes and error-proofing systems, and making significant progress toward single-digit ppm and even meeting the goal of zero defects to your customers.
- Reduces variation in production.
- Improves quality.
- Improves and maintains discipline.
- Reduces scrap and eliminates waste.
- Reduces customer rejections.
- Improves discipline and communication.
- Increases employee participation.
- Improves overall quality and cuts costs.
- Stops production problems from becoming rejections.

The business impact of LPA is the reduction of variation, which is waste which translates into profitability. A layered process audit is an ongoing chain of simple verification checks that ensure a defined process is followed correctly. It is a powerful management tool that can improve safety, quality, and cost savings by amplifying problem-solving systems and making continuous improvement almost routine. Through observation, evaluation, and conversations on the manufacturing floor, these checks ensure key work steps are performed properly.

Furthermore, LPA interactions are also an excellent way for managers to show respect for frontline workers. Some essential characteristics of LPA that have a direct business impact are:

- Audits are done on the *Gemba* (source). There are no filters to management.
- Management must take ownership of the LPA process.
- Auditors must identify and ask the right questions.
- All management layers, top to bottom, must participate.
- Immediate containment of nonconformances found.
- Continual improvement must be included in the process
- Shift, daily, weekly scheduled and performed audits. Survey data shows excessive process input variation is the single largest cause of manufacturing quality problems. This often results from a failure to reinforce process corrective actions.

Error-Proofing Audit

Error-proofing audits are generally confirmation audits. This means a nonconformance was identified, a corrective action was recommended, and now the auditor checks whether or not the action taken was effective. These audits are simply an organized group of questions designed to examine a device or process after the process has been error proofed. Therefore, they are focused only on areas in the manufacturing process where deviation represents a high risk for producing defective products. For an error-proofing audit to be effective, it must integrate action, analysis, and improvements. If an auditor finds a nonconformance while performing this type of audit, that auditor should not only record their finding but also take immediate initial corrective action to ensure defective products do not get out the door. Information about the finding should be recorded and readily available to management for later analysis. With a good system for recording and reporting audit information, the error-proofing audit records provide an excellent tool for troubleshooting problem areas and identifying other similar places within the organization for improvement. If an error-proofing audit is to be conducted, a typical partial form is shown in Table 9.13.

TABLE 9.13
A Typical Partial Error Proofing Layered Audit

Error Proofing Audit ***Only qualified personnel can perform error proofing verification audits***										
Department Audited:									Date:	
Auditor's Name:									Shift:	
Note: All questions are to be answered with a "Y" or "N" for yes and no, respectively, under the corresponding day's box. ALL non-conformances *must* have corrective action recorded under the specified column in this form with the date (MM/DD/YY)										
Item no.	These items must be checked every shift	Audit area	M	T	W	T	F	S	S	Corrective action
Section 1: General										
1	Are operators aware of the function and importance of any error proofing devices at their station?									
2	Are all error proofing verification parts identified and maintained with due care?									
3	Are all error proofing devices verified per work instructions? Are the verification results recorded?									
4	When error proofing devices fail verification checks, are the proper employees notified? Is the contact documented? Is there any follow up?									

(Continued)

TABLE 9.13 (*Continued*)
A Typical Partial Error Proofing Layered Audit

5	Is there a reaction plan to direct operators when error proofing devices are not functioning? That is, how will the known defect or potential defect be detected in order to protect the customer in the absence of automatic error detection?									
6	Are all operators performing error proofing verification checks trained to do so?									
Section 2: Work Cell Specific										
1	This area of verification will be specific and unique for every work cell containing error proofing.									

GENERAL COMMENTS ABOUT THE AUDITS

By far the most important item in any audit is the element of surprise. Observations (walk the line) allow auditors to observe employees completing specific tasks in the company's work environment. Auditors can best use this process by showing up unannounced and conducting an observational review of the company's operations. This allows auditors to avoid managers and employees who will operate under the company's procedures for the planned audit day. Auditors need a true and unfiltered audit observation to ensure that employees properly fill out paperwork and follow all operational standards.

Many organizations require employees to fill out documentation of their activities in the working environment. Auditors review these forms to see which employees are filling out paperwork and the times or dates of the paperwork. This holds employees accountable for their actions, especially operational managers who must uphold the company's safety or standard operating procedures. While it can be difficult to determine if the paperwork is fraudulent, combining this process with employee interviews can help in the review of this information.

Consistency is also important. Audits must be carried out on a regular basis. While LPAs seem to use up a lot of human resources – since many employees must engage in audits – they ultimately reduce company waste by identifying processes that are wasteful or a threat to the overall success of a project. However, the most valuable aspect, especially for an LPA, is the fact that it allows management to

- Verify critical processes for conformance to requirements.
- Give management a reason to be on the floor.
- Ensure problem-solving solutions (corrective actions) stay in place and are monitored.
- Identify problems before the issue reaches the customer.

Yet another item necessary for any audit is the checklist. During any audit preparation, auditors fill out checklists to acquire the data necessary to fully understand the processes within the company. Depending on the audit, many levels of the company must assess the audit checklist to ensure that the checklist is perfected. Also, checklists should be clear and specific so that different levels of management will come to the same conclusions when answering compliance checklists. Typical things to remember when constructing a checklist are:

- Management must own the process, so ask appropriate and applicable questions.
- Identify and ask the right questions for the process being audited. The fact that management is involved establishes accountability and encourages operator feedback at the source (Gemba). Of course, the starting point should be the highest priority for either a quality issue or risk for the customer.
- Participation spans all management layers. Make sure the questions are focused on the appropriate level. As the level increases, the questions become more strategic and policy-oriented.
- Nonconformances lead to immediate containment. Make sure that there is a corrective action for each nonconformance. Quarantine alone is not a complete action. A typical form for keeping track of nonconformances in an LPA is shown in Table 9.14.
- Include a process for continuous improvement. Make sure that CI is the result of your findings, not busy work.
- Schedule and perform audits regularly. This is a very important consideration, especially for the layered audit. Since the upper-level management staff that conduct audits do not always have a lot of time on their hands, audits usually need to be quick and simple. The management occasionally observes employees to see if they are complying with policies and procedures. The upper-level management audits can be somewhat unpredictable in when they perform audits, but lower-level auditors typically audit the same series of items in their department every day. Typical participation may be as follows:
    - Plant manager once a week.
    - Area managers once a week.
    - Production supervisor once a shift.

TABLE 9.14
A Typical Form for Tracking Non-Conformances in an LPA

Layered Audit Results						
Items NOT in compliance						
Item	Number of occurrences	M	T	W	T	F

PREPARATION OF AN AUDIT

Hopefully by now the reader has recognized that audits are of paramount importance for any organization, because they uncover nonconformances of "the" system used, but more importantly, they make sure that the VOC is heard and is part of the ongoing improvement within the organization. Audits can be based on international standards, customer requirements, government regulations, and or internal improvement protocols. In Table 9.15, we have identified several areas of concern, and we split them into three categories. The first one is the required items, the second is the criteria of the requirements, and the third is the verification possibilities. Obviously, this is not an exhaustive list and certainly not a universal characterization, but it can serve as a model for any audit in any organization. In any case, the items identified here are a demonstration of what an organization must be able to accomplish with an audit. It is a road path for preparing for excellence.

TABLE 9.15
An Overview of Criteria for Preparation of an Audit

Required	Criteria of Requirements	Verification Possibilities
Fast Response		
Daily leadership meeting held with cross-functional, multilevel attendees to address significant external and internal concerns.	1. There is a daily Fast Response (FR) meeting with cross-functional attendees and led by manufacturing. 2. The FR meeting is a communication meeting and a precise timing is respected. 3. All the significant external and internal issues are addressed. 4. Natural owners are assigned to problems, next report-out date is assigned. 5. Take into account the production forecast and the quantity produced. 6. Safety and near miss accident issues are reported out.	Attend FR meeting. Observe: • Lead by manufacturing with cross-functional attendees. • How leader controls the FR meeting (keep timing max 10–20 minutes, focus on subject, not going into the details...). • Environment is suitable (everyone can hear and see the meeting), • How issues are reported out. • Problem Solving report format is used to report out and document the status of the issue.
Fast Response Board tracks all major concerns with appropriate timing and exit criteria.	1. Fast Response board is used for tracking of issues, being updated before FR meeting (board could be a dry erase board, laminated poster, Excel sheet projected by Dell projector or something similar, etc.) 2. Method of communicating problems to all Key Stakeholders is defined (*e.g.*, Quality Alert, Temporary Work Instructions, By-Pass Procedure, *etc.*)	• Prior to the audit check, last customer complaints and focus on open ones. • Prior to the FR meeting, ask if there are any significant internal issues. • Check the board if it contains above above-described external and internal issues.

(Continued)

TABLE 9.15 (*Continued*)
An Overview of Criteria for Preparation of an Audit

Required	Criteria of Requirements	Verification Possibilities
	3. Exit criteria represent the core 6 Steps of problem solving (1. Define, 2. Contain, 3. Root cause, 4. Correct, 5. Validate, 6. Institutionalize), including timing for each of the exit criteria. 4. All the exit criteria are identified (Red, Yellow, Green), red items have a planned date to go green with next steps. Problems are not closed until all criteria are met. Overall status represents the worst condition, or the overall planned timing has been exceeded.	• Follow an issue from FR Tracking Board through the exit criteria, confirming actions are in place and all documentation has been updated. • Check a few statuses if they are rated well based on their timing, judge a few N/A items whether they do not need to close the issue.
System is in place to ensure fast response to operator's concerns and information flow between shifts.	1. System in place to support responding to the concerns of the operator. 2. Escalation process is defined in order to ensure that problems are quickly communicated to people who can help. 3. If there is more than one shift, all the information that can affect to next shifts have to be passed across and documented. 4. Manufacturing leadership reviews shift book daily to verify proper containment or corrective action done, decides further escalation to FR meeting.	• Ask operators how they can escalate their issues. • Andon system or similar in place, if applicable. • Test Andon's function and response to request (Light Boards, lamps, or audio signals work, support arrives soon). • Participate in shift change, check shift book, and information shared.
A defined process for Problem Solving, including a standard for documenting the tools used for root cause identification.	1. Standard process (practical problem-solving report (PPSR), 8D or equivalent) used with a format that follows the core "6 Steps" of problem solving (1. Define, 2. Contain, 3. Root Cause, 4. Correct, 5. Validate, 6. Institutionalize). 2. Analysis process is developed and applied to all the customer returned parts. 3. Problem-solving forms used across the plant for internal, customer, and supplier issues.	• Prior to the audit check last customer complaints focus issues where root cause found and corrective action implemented. • Verify that problem solving used efficiently, all the core "6 steps" applied, especially that real main root cause found and action implemented against the root cause.

(Continued)

TABLE 9.15 (*Continued*)
An Overview of Criteria for Preparation of an Audit

Required	Criteria of Requirements	Verification Possibilities
	4. Evidence of cross-functional team approach and team members got the necessary training about problem-solving methods applied. 5. Tools for identifying root cause are applied (GM's 7 diamonds, Ford's G8D, 3X5 why to correct systemic issues, Fishbone Diagram, *etc.*).	• If no customer complaint issued, verify via an internal or sub-supplier issue. • Check a Drill Deep (5 whys), main systematic root causes found.
A system to capture and institutionalize lessons learned.	1. Electronic form or database is used to document Lessons Learned. 2. Method that assures implementation supported by evidence of review dates, distribution lists, or posted Lessons Learned, as well as reviewed by Leadership. 3. Institutionalization is applied to all the issues where problem solving done. 4. Completed Lessons Learned information, which is easily retrievable by all who need the information. (*e.g.*, Master FMEA, APQP Program check list reviews). 5. A process to collect suggestions from personnel is put in place and contribute to Lessons Learned and process improvement.	• Ask people for examples how they are using Lessons Learned system. • Check 6th step of problem solving (Institutionalize) via examples of point FR4. Check Drill Wide or D7 (8th step) of 8D of last customer complaints.
Targets are defined and followed to ensure effectiveness of fast reaction for external and internal issues.	1. Tracking of safety-related metrics (*e.g.*, accident without lost working days, distinguishing lost time accident & no lost time accident, *etc.*). Lead time to solve safety issues. 2. Quality Q or other method to track FR performance is posted and up to date. 3. Tracking of the total amount of significant internal and customer complaints based on their trend and statuses (not responded, open, or in delay and closed in time) and average closing timing.	• Prior to audit check number of last 12-month PRRs. If higher than 24, do not count line accumulation ones, or trend is significantly negative special PRR reduction team has to be established. • Check last customer complaints whether due dates kept. If not reasons for delay, actions need to be addressed. • Check red items percentage, evaluate actions addressed to eliminate roadblocks.

(Continued)

TABLE 9.15 (*Continued*)
An Overview of Criteria for Preparation of an Audit

Required	Criteria of Requirements	Verification Possibilities
	4. Tracking of the repetitive issues occurred after corrective action implemented. 5. Continuous improvement data for line stoppage/downtime from Andon. 6. Location with a high level of customer complaints (>24 complaints in the last 12 months) shall have a special team created to work on complaints reductions – performance shall be tracked at FR meeting on open issues list.	• Evidence of periodical review of average closing time for each exit criteria and set action plan for any deviation.
Control of Nonconforming Product		
Traceability rules are applied according to the class of traceability of the finished product, and FIFO is kept.	1. Development of processes integrates traceability requirement (including tier parts) agreed with customer and applied at each stage of the manufacturing process. 2. Quality status of the product (physical or data-processing) and the last operation performed are identified in entire process. 3. Equipment and stock areas are designed to facilitate FIFO and allow to identify easily the noncompliance with FIFO. 4. When FIFO cannot be strictly respected (derivative flows), rules are defined to minimize the disturbances (*e.g.*, stock rotations). 5. "Quantitative" approach is taken into account to determine theoretical suspected batch size. "Precision Calculations" (Dilution calculation) are carried out, and rules/procedures are defined to comply threshold of Dilution.	• Check parts in process and suspected material that last operation performed is identified. • When traceability is required check it via a part. If you have unitary traceability on a parameter or a component: choose a finished product and ask for the component / process "value" associated with the product. • Assure that layered process audit (LPA) check sheet is assuring traceability application. • Precision calculation: Check how many finished products must be verified if an incoming part, a process parameter, or a rework process failed.

(*Continued*)

TABLE 9.15 (*Continued*)
An Overview of Criteria for Preparation of an Audit

Required	Criteria of Requirements	Verification Possibilities
Nonconforming and suspect material are identified and segregated in order to prevent them from unintended use.	1. Consistent tagging and Visual management (foot print, labels, *etc.*) procedure is defined and kept in entire organization to ensure that identification and handling of not OK or suspected material is in place to avoid mixing with normal parts. 2. If random checks (*e.g.*, sample check, product audit) find not Ok (NOK) part, it is ensured that all the parts produced from last known good parts are handled as suspected. 3. If red tags are used for both scrap and suspect material, tag must have disposition. 4. NOK parts are segregated, recorded, and their storage is managed, including Quarantine area which has authorized access, and quantities in quarantine are controlled.	• Check in shop floor that identification tagging system is used in all areas including documentation (check incoming, working stations, control station, rework) and including visual management like foot print, color coding, labels *etc.* • Ask operators that everyone in shop floor is understanding color coding used inside the organization. • Scrap boxes size should match with part size. • Verify quarantine, access is defined, quantity is controlled. Verify that avoidance of mixing NOK part ensure via layout of workplace, handling and storing of NC parts.
Alert is issued for significant issues to ensure communication and action for stakeholders.	1. Alert process is defined on the basis of the severity level of nonconformity and applied for both external and internal issues once problem occurs at least as per FR issues. 2. Alert includes minimum problem definition, the standard deviation from standard, tasks, time line. 3. Escalation process is put in place with internal/external distribution and notification (Contact list for customer and tier suppliers exits). 4. The standard defines the criteria of withdrawal of posting.	• Review - FR board and check open issues have quality alert posted • Check Quality Alert is clear, understood by team, and posted in stations needed (Operation, Quality) • Customer and tier supplier contact list are available

(Continued)

TABLE 9.15 (*Continued*)
An Overview of Criteria for Preparation of an Audit

Required	Criteria of Requirements	Verification Possibilities
Containment is segregated and standardized, all the potential quantities and locations of suspected material are identified.	1. Containment worksheet or equivalent contains all the potential locations and quantities, including physical quantities and status (OK/NOK). 2. Containment area is separated from production line, standardized work is applied. 3. Containment process systematically includes securing of stock and pipeline and application of immediate countermeasures to guarantee conformity of production and deliveries, break point identified to customer. 4. Containment actions are defined for each customer issue and verified in order to prevent further defects, breakpoint are documented and communicated to all involved parties. 5. Exit criteria are defined to stop countermeasures.	• Verify that Containment Worksheet is used before initiating any containment. • Check Containment Worksheet contains all the potential locations. • Check that countermeasures are put in place for each alert. • Ask team member about containment and its rules. • Ask one team member who performed containment in any FR issue, how he did the containment, and how he filled the Containment Worksheet.
Rework or repair is standardized, performed only with necessary authorization, and process of reintroducing parts back to line	1. Prohibited or authorized rework operation clearly defined. Authorized rework operations must be part of part/process approval. Parts waiting for rework have to be handled as suspected parts. 2. Failure modes of rework are considered in PFMEA, identified in Process Flow and control plan with its reintroduction at or prior of removal point. 3. Reintroduction of reworked part includes all downstream checks and color coding in order to ensure that all control plan inspections and tests are performed. 4. If a rework needs to be applied that is not part of approved process, authorization (if needed includes customer) and releasing responsibility are defined.	• Rework is part of Process Flow. • Potential failure modes of rework are detailed in PFMEA. • Check rework station, standardized work is applied. • Check if team members understand rework identification process and follow one rework part, how handled, identified, and how reintroduced back to line.

(Continued)

TABLE 9.15 (*Continued*)
An Overview of Criteria for Preparation of an Audit

Required	Criteria of Requirements	Verification Possibilities
	5. Re-use of components is considered as a rework operation. The component re-used must be traced on the finished product. 6. Each reworked part is traceable via an identifier, documented serial number etc.	
Parts that have deviations but approved by customer are managed; traceability of concerned products is guaranteed.	1. All the parts that have deviations are handled as suspect parts. 2. Deviation is approved according to customer requirement. Deviations are granted for a limited period or a defined quantity of parts. 3. Traceability of the parts delivered under deviation is guaranteed. Manufacturing batches are identified. 4. Impact of deviation on process and product is analyzed, and based on its result all the responsible parties must be included in communication and decision about usability of deviated parts.	• Site history of requests for deviations. • Procedures and forms used on-site. • Check PPAP or appropriate customer-specific requirement. • Check "Request to deliver non-conform product" for PSA and initial samples are available for deviations. • Multi-skilled approach for the decision to initiate deviation request. • Evidence of traceability.
Targets are defined and followed to ensure effectiveness of fast reaction for external and internal issues.	1. Dilution calculation: affected batch size vs. containment worksheet in case of internal or customer complaint. 2. Performance metrics such as scrap rate (internal ppm), rework percentage, first time quality (FTQ), voice of the customer (VOC), *etc.*, are established and followed at all levels of the operation. 3. Tracking of customer complaints caused by rework, best practice violation, and known NOK parts shipped. 4. Costs of poor quality (including indirect costs: sorts, *etc.*). 5. Tracking of number of containment at customer and in-house. 6. Tracking time in containment (long-lasting)	• Prior to audit check Controlled Shippings and customer complaints link to control on non-conforming product. • Verify that containment driven back to the source of error. • Verification of countermeasure action.

(Continued)

TABLE 9.15 *(Continued)*
An Overview of Criteria for Preparation of an Audit

Required	Criteria of Requirements	Verification Possibilities
Mistake Proofing Verification		
A system in place focuses on Building Quality in Station through prevention, detection, and containment of abnormalities. [The order here is very important.]	1. Verification Station (VS) strategy is defined on procedure/instruction level, contains selection of VS place, guidelines for alarm limit, and definition of escalation route. 2. Characteristics have to be checked, 100% are defined and Verification Station needs to be developed according to standardized work requirements. 3. Customer complaints have to always be covered in verification station. 4. Permanent verification station is implemented based on customer requirement. 5. Exit criteria are defined to remove non-permanent Verification Station and approved by Quality. 6. Impact of VS on capacity of the line is verified.	• Ask for set up a Verification Station for a theoretical problem and check that conditions are defined to establish VS in short time. • Check a verification station, is clearly identified, developed according to standardized work: instruction developed, layout defined to avoid bypass and mixing of parts, training and necessary certification done. • Check for customer requirements compliance. • For pre-supplier agreement, suppliers' final inspection is implemented as Verification Station. • If needed, VS capacity confirmation via "limited" R@R.
Alarm system is defined and performance tracked for defects entering Verification Station (VS).	1. Results of controls are recorded and posted at or near VS. A real-time follow-up of results is applied. 2. Escalation procedure is defined and followed when the Alarm limit reached. 3. Issues are documented and countermeasures are applied immediate. 4. The decision criteria to stop production are established.	• Tally Sheet posted at or near the Verification Station, filled in properly. • Ask VS operator about escalation. When and who to call, who responded and when. • Check results back, proper escalation was done when alarm limit reached. • Alarm limits are reasonable: *e.g.*, for customer complaint, specific productivity tool, *i.e.*, workflow, containment system.

(Continued)

TABLE 9.15 (*Continued*)
An Overview of Criteria for Preparation of an Audit

Required	Criteria of Requirements	Verification Possibilities
Problem problem-solving process is applied, communication is in place for defects leaving station.	1. Problem resolution management is quickly applied directly at the workstation. according to escalation route, following problem solving process. 2. When alarm limit reached, an upstream reaction process is defined. 3. Leadership reviews verification station results and follows up on action plan, if needed, institutionalize/read across its results. 4. In case of lack of detection of nonconformance part, verification station controls must be re-evaluated and improved.	• Check problem solving was applied, and corrective actions were defined against the main root cause (not re-training of operators). • Verification station is owned by management.
The Error Proofing (EP) devices are identified, managed, and regularly verified.	1. A master list of error-proofing devices is available, including description of verification (self or with master samples). 2. Each error-proofing device/system and its master samples are clearly identified and managed. 3. Error-proofing devices are verified regularly with master sample at least once per day, but at all part number changes and start-ups (including after a significant production stop). 4. Verification documented with operator sign-off. 5. When it is applicable, Gage R&R has been conducted to confirm EP efficiency.	• Participate at an Error Proofing verification, check process kept and documented well. • Identification of EP devices on shop floor & coherence with the list. • Records of verification (control plan, start-up work instructions). • Check identification, conservation, easy access, calibration of master samples. • Work Instructions for verification. • Is Gage R&R result less than 10%?
In the event of Error Proofing devices malfunction, the suspect products are managed.	1. All the EP failures have to be documented and the reaction plan includes who is notified and actions to be taken. 2. All the parts produced since the last OK verification have to be handled as suspected material and apply containment.	• Ask people who make EP verification about his/her responsibility in case of EP failure and escalation process. • Check back records that containment was done and completed for all the EP failures.

(Continued)

TABLE 9.15 (*Continued*)
An Overview of Criteria for Preparation of an Audit

Required	Criteria of Requirements	Verification Possibilities
Reaction to failures, corrective actions and reverification are documented.	3. Corrective actions to fix Error Proofing device failure are documented. 4. In the event of error proofing devices malfunctioning or unavailability, production is stopped until capable substitution processes/control are identified and handled as by-pass process.	
Targets are defined and followed to evaluate Verification Station effectiveness.	1. Performance Graphic, Q chart, or report card (*i.e.*, calendar days-red/ green, I-chart etc.) showing feedback from downstream and upstream processes. 2. FTQ or internal scrap metrics showing improvement trend, reduction of events/defects over time. 3. Verification station applied flexible, alarm limits are reviewed continuously. 4. Tracking of error proofing failures.	• Prior to audit check customer complaints caused by failed error proofing or no detection on Verification Station. • Any long-lasting processes or temporally verification station. Actions to close them. • Review charts, verify action brought expected result. • Check how often the alarm limit is reached: no alarm = no improvement. • Check that verification station(s) selected efficiently: 1. *Review data*: customer complaints (any major or repetitive issue), FTQ results, high severity #s or RPN items from PFMEA, process capability data. 2. Based on data reviewed, evaluate if Verification Station(s) implemented to right place or there is a need to implement a new one(s). • Verify that error proofing verification frequencies are reasonable.
Standardized Work		
The workplace environment are safe and ergonomic.	1. Relevant safety standards are applied for each workplace. 2. Design of general and workstation layout considers safety aspects to avoid potential safety risks.	• Check on shop floor potential safety issues, *e.g.*, hidden corners, potential accidents, pedestrian way, color coding on floor, noise, temperature etc.

(Continued)

TABLE 9.15 (*Continued*)
An Overview of Criteria for Preparation of an Audit

Required	Criteria of Requirements	Verification Possibilities
	3. Rules related to ergonomics of workstations are defined and applied for each workstation design. They take into account the environmental conditions. 4. Operators are involved in workstation design (ergonomic work stream) and major player in the qualification process, which always includes safety and ergonomic aspects.	• On the shop floor, appreciate the level of light, the temperature (cold/hot), the level of noise, the loads carried by operators, the level of work (hands up, *etc.*). • Look for result of ergonomics evaluation. • Look at a "painful" workstation. Is there a specific verify its action plan?
The workplace is clean and orderly arranged in order to contribute to higher quality and optimization.	1. A process to define layout, including internal stock or buffer, is available. 2. Organization of workplaces are suitable to ensure compliance with the FIFO, including repackaging operations. 3. Systematic approach for all the workplace organizations (like 6S) is implemented and maintained. 4. Visualization is used on the shop floor 5. A continual improvement and optimization / waste elimination process is in place related to workplace organization.	• Check workplace organization and visualization at several different places (incoming/ storage area, work stations, maintenance room). • 6S audit records and verify actions implemented for findings. • Layout is in coherence with the workstation. • Method for waste elimination is applied (VSM, 7 Waste, *etc.*) and periodical review minutes are available.
Working instructions are standardized and available for each workstation.	1. Standardized Work Instruction (SWI) are available for each operation. 2. SWI covers the whole of the produced references, and line balancing means leveling of the workload across all workstations. 3. Revision history is traced back for process/product changes. 4. SWI is developed by cross-functional team and includes at minimum: • Work elements including quality controls and their sequence, • Operator movement with sketch of work flow,	• Work instructions at different workplaces (manufacturing, quality control, material handling/logistic), check: 1. Easy availability 2. Compare instructions to work performed by operators. Observe 3 full cycles of the job in station and verify that the Major Steps, Key Points are followed, Reason is understood

(Continued)

TABLE 9.15 (*Continued*)
An Overview of Criteria for Preparation of an Audit

Required	Criteria of Requirements	Verification Possibilities
	• Takt time and overall cycle time, • Standard in-process stock, • Required PPE and safety requirement if applicable, • Support description with pictures, sketches, and images, • Reference to product/process/control standards. 5. For each work element, instruction describes: • Major Steps (What), • Key Points (How), • Reason (Why). 6. SWI is posted near or at the operation.	3. Controls listed in Control Plan are added to SWI, efficient time allocated for quality checks. • Try to perform a manual operation based on SWI, check all necessary information, hint, key points... described to perform operation. • Ask few operators to explain SWI. Do operators understand it? • Various balances are managed for different planned production outputs and product mix (*e.g.*, modification of line speed).
The operations to validate the start of production are described and applied.	1. Start-up instructions are available (can be covered in SWI or checklist) for manufacturing operation and tool change/set up. 2. A Start up process is applied at beginning of shift, part number change, after significant production stop (planned or unplanned) and documented. 3. Start up instruction includes a list of the checking tasks to be carried out and recorded: • availability and functionality of all the manufacturing and control equipment, error proofing and personal protective equipment (PPE), • process/product parameters with tolerance limits, • availability of components and materials, • environmental conditions at the workstation (cleanliness, lighting…).	• Take 2–3 Start-up instructions for automatic equipment and verify content. • Ask a set-up person about roles of set up. • Evidence of traceability between parts produced and 1st off part. • Check back records for startup activities and verify: • exact date of start-up documented • set-up parameters recorded and within tolerance • first-off parts result • in case of any deviation action initiated and result verified.

(Continued)

TABLE 9.15 (*Continued*)
An Overview of Criteria for Preparation of an Audit

Required	Criteria of Requirements	Verification Possibilities
	4. All activities/operations to be carried out prior to tool installation are defined (*e.g.*, pre-heating of tool). 5. Start up instruction defines first-off parts validation. Traceability to be ensured till validation completed, result is documented.	
Reference parts and boundary samples are available and managed in order to confirm inspections.	1. Responsibilities for definition of samples are established (including customer if required). At boundary samples tolerances for each characteristic and decision criteria are clearly established. 2. Samples (some cases they can be replaced by photos) are clearly identified and in accordance with latest design and approval status. 3. Usage of samples is described/ referred in instructions, used in training process. 4. Samples are periodically reviewed and its result is documented based on acceptance criteria and customer feedback/complaint.	• Verify that boundary samples are available for operators. • Check that samples represent typical failures. • Ask operators when they using samples. • Verify sample storage condition. • Evidence that samples used for training. • They have easy access in area where they are used, storage preserves original condition.
The capability of the control devices is checked prior to using them in production process, and it is periodically verified.	1. All the gauges are periodically calibrated and recorded according to procedures, ensured that only calibrated gauges are in use. 2. If calibration is performed in-house, the necessary skilled staff, equipment, and facility are available; if outsourced, external laboratory accreditation is verified. 3. The capability of the measurement systems/equipment (Gage R&R) is periodically reviewed according to procedures. 4. The acceptance criteria are defined for calibration and capability.	• Check several gages in different area (production, lab, incoming, storage) for: 1. identification, 2. calibration status and their record, 3. proper usage and storage (ask operators about usage and handling of gages, are aware about risks of damaged gage). • List of gages contains identification and calibration period. • Check schedule for calibration.

(Continued)

TABLE 9.15 (*Continued*)
An Overview of Criteria for Preparation of an Audit

Required	Criteria of Requirements	Verification Possibilities
	5. For each deviation/non-conformity or equipment exceeding calibration due date, containment and corrective actions are defined and validated, and followed by Quality Manager (apply handling of non-conformance). 6. Standardized Work is applied for gauge instructions.	• Verify a work instruction for a gage (see the acceptance criteria). • Gage R&R are been conducted to calibrate operators from shift to shift / line to line.
Target defined and followed to optimize processes	1. Tracking of external and internal issues created by not well-defined working instructions. 2. Layered audit result related to non-compliance of Standardized Work. 3. Direct labor efficiency (ratio of the real number of parts produced against the theoretical number to be produced during the opening time). 4. Cycle times leveling. 5. Capability and calibration follow-up indicators (*e.g.*, number of late calibrations). 6. Tracking of downtime and/or scrap rate caused by setup. 7. Ergonomics evaluation rate.	• Prior to audit, check customer complaints where the root cause is linked to Standardized Work (working instruction not detailed, wrong setup, *etc.*).
Training		
Training process is defined. Organization, methods, skills, and facilities are available for training activity.	1. The training process covers the entire staff of manufacturing, logistics, and the support functions, including temporary employees. 2. Internal trainers are identified and qualified to ensure that only certified trainers train. 3. A graduating step approach is developed and applied for each training activity, such as the 4 step: (a). Prepare, (b) Demonstrate, (c) Try-out performance, (d) Follow-up. 4. Dedicated infrastructure (outside of the process) for hands-on training simulating production conditions, if possible.	• Verify at different workplaces whether their operators got the training from certified trainer. • Record of Internal Qualified Trainers. • Ask operators how they were trained. What specific training did they receive? – Rules for graduation process is described in training procedure.

(Continued)

TABLE 9.15 (*Continued*)
An Overview of Criteria for Preparation of an Audit

Required	Criteria of Requirements	Verification Possibilities
Training needs for safety are identified and training is performed for all the relevant people.	1. A safety policy is defined and communicated to all staff. A safety handbook is established and regularly updated in coherence with the policy. 2. An organization is defined according to policy and local legislation with suitable equipment and qualified people. 3. Training needs for safety are defined based on the risk assessment linked to the industrial domain and specific workstation. 4. Based on safety training needs, all the relevant staff are trained in two levels: a. Basic training about general safety rules, b. individual job training, if applicable. 5. Visitors' safety is managed; Standards and Equipment are available (quick training, suitable safety devices are at disposal, *i.e.*, safety glasses, hearing plugs, appropriate gloves, covers for static electricity, *etc.*).	• Safety manual and policy. • New employee training covers safety requirements. • Ask operators, who works on station where safety requirement established, about awareness of safety rules. • Organization chart: Responsibility for the safety is defined. • Facilities on site: first aid station, *etc.* • Safety standards are kept, *e.g.*, Personal Protective Equipment (PPE), circulation on the shop floor, risk related to process (stamping, welding, assembly, *etc.*). • Signs, posters on the shopfloor, line marking, and behavior of logistic employees.
Basic and individual job training needs are identified, their content is formalized, and all the trainings are recorded.	1. Basic training is given to each new employee, which includes: • New hire orientation, • Knowledge and respect of the product (including management of key characteristics), • Perform proper record keeping (production/quality), • Understand workplace organization responsibility, • Quality requirements (containment, red bins, Andon, *etc.*), • Escalation procedure (are the right people notified during a problem).	• Check if the Individual Job Training has all training content need to perform the task. • Check Individual Job Training records for different job element (operator at a working station, quality controller, team leader, a support function) • Ask team member to explain the documentation regarding quality, workplace organization, escalation process. • Verify few operators whether following Standardized Work Instructions and know the quality and productivity requirements.

(Continued)

TABLE 9.15 (*Continued*)
An Overview of Criteria for Preparation of an Audit

Required	Criteria of Requirements	Verification Possibilities
	2. Individual job training is documented for each employee, showing training needs, skill or knowledge level of the job, who trained them, timing of operator sign-off. 3. Individual job training is held in case of a change in training content or need (*e.g.*, Working Instruction change). Traceability for the latest change is ensured. 4. Documentation, scheduling, and tracking forms for employee refresher training if they have not performed that job within a specific duration. 5. All the training activities are documented, and records are available and easily retrieved.	• Check few working stations and verify that training records are available to the latest working instruction level. • Verify if there are employees that go back to the operation after specific period out of operation without refresh training.
Operator qualification process for each job position and workplace is applied, including re-qualification if needed to ensure that only qualified people perform the job.	1. Qualification levels are established. For each one of them, measurable criteria are defined. 2. A calibration process is in place for all the people making check operations, where the result depends on a subjective decision. 3. Flexibility Chart or equivalent posted at all operations or work areas, which • contains number of qualified people per workstation, as well as workstations per person are targeted; associated action plans are implemented, • indicates the steps in training and skill qualification level achieved for each job, • has been updated. 4. Criteria to revise qualification level are defined; they take into account the operational results at the specific workstation, the result of the layered audit, time off job, *etc*.	• Check several operations where result depends on Subjective Decision. Check the record of calibration process, frequency, results, and action plan. • Chart showing cross training/certification level in a cell or work area such as a flexibility chart. • Look for a job rotation plan or log. How often does team rotate? • The number of Team Members certified per station should support the Job Rotation Plan. • Check if the training procedure describe the Re-Qualification process • Check if a Re-Qualification Process (Employee Performance Review) is in place. Evaluate if an action plan was generated in case of Low Performance.

(Continued)

TABLE 9.15 (*Continued*)
An Overview of Criteria for Preparation of an Audit

Required	Criteria of Requirements	Verification Possibilities
	5. If re-qualification failed, actions are implemented to reach the required qualification level again, including re-assessment or degrade qualification level.	
A process of motivation (individual recognition / reward) is implemented. The policy is defined and managed.	1. Process in place to recognize employee's achievements. 2. Individual objectives are in line with the site targets and are regularly reviewed, evaluated with feedback to the employees. Individual interviews are systematic for the whole staff. 3. Process in place to collect and address employee grievances with short- and medium-term improvement actions established.	• Check how the leadership is promoting the recognition (*e.g.*, 7 Recognition Success Factors – see glossary). • What kind of system does the supplier use to receive feedback from the employees? • Check if the action plan regarding employee grievances is "on track." • Verify if the individual objectives were deployed from the Business Plan.
Metrics are defined and followed to ensure effectiveness of training process.	1. Flexibility target (number of qualifications per employee & number of employees qualified per workstation). 2. Layered audit results related to people awareness of procedures (*e.g.*, standardized work, safety requirements, material handling, workplace organization, *etc.*). 3. Number of re-training needed in the same subject (failure of first training). 4. Training assessment by the trainee. 5. Internal and external issues due to the lack of training. 6. Employee satisfaction survey.	• During audit check how team members follow general procedures (*e.g.*, handling of non-conforming material), as a confirmation about training effectiveness. • Check all the findings found during audit that can be linked to a lack of training. (The root cause of deviation is a lack of knowledge.) It affects to effectiveness of training. • Does supplier measure the turnover in their departments (fluctuation) - reason why?
Audits		
A generic Layered Process Audit (LPA) is established.	1. Written procedure: Defining the rules of LPA. 2. Layered audit check sheet: developed and applied for all manufacturing and material handling (logistic) areas. LPA Check Sheet shall define what to "Look For."	Verify if the LPA procedure define: • Frequency • who shall perform the LPA • how to conduct the LPA (standard method) • how to record and treat issues

(Continued)

TABLE 9.15 (*Continued*)
An Overview of Criteria for Preparation of an Audit

Required	Criteria of Requirements	Verification Possibilities
	3. LPA Check Sheet comprises the following elements at minimum: • *Safety and Ergonomic items* • proper safety practices and PPE are being followed. • *Manufacturing/Material Handling (logistic) items* • Standardized Work is being strictly followed, • Workplace organization standards are maintained, proper tools, gauges, and materials are available and used, • Operators work according to Flexibility Chart, • Standard in-Process Stock is according to Standardized Work, • FIFO is being followed. • *Quality Focused* • Required quality controls completed and documented acc. to Process CP, • Error Proofing verification done and documented properly. 4. Training of auditors: using a standard method for LPA (choose the workstation, conduct the audit, give feedback, document results, and conduct follow-up). 5. LPA is applied for standard processes of supporting functions, and Internal Process-specific audits are performed (*i.e.*, Process/ Commodity Specific Audit, CQI audits, *etc.*).	Examples of fields in the LPA check sheet form: *Safety* • PPE: the team member is using all the posted Personal Protective Equipment *Manufacturing/Material Handling* • Work Instructions (*e.g.*, Standard Operation Sheet, Job Element Sheet). • Proper tools, gages and materials available and used. *Quality Focused* • Ensure control of significant process elements which can impact areas such as Customer Satisfaction, PPM, Warranty. • Ensure control of high risk elements including: Operator/ Process Sensitive Operations, Key Process Control Operations/ Checks, Mandatory Assembly Sequence Operations. • Verify that auditor understood the questions, LPA checklist filled in properly: Follow a team/ group leader in an LPA. Verify if they use the Standard Method.
LPA, which covers the whole operational activities, is carried out and owned by manufacturing.	1. Each workstation is audited and the schedule frequency is defined to ensure: • covering all the shifts, • each operation is audited a minimum once per month, • containment activities (*e.g.*, sorting, Controlled Shipping, specific processes, *etc.*) are covered by LPA.	• In the shop floor, select a workstation and verify how the workstation is audited by LPA: • frequency is according to the plan, • once a problem is identified, how the team member is informed, • check if safety issues are detected by LPA,

(Continued)

TABLE 9.15 (*Continued*)
An Overview of Criteria for Preparation of an Audit

Required	Criteria of Requirements	Verification Possibilities
	2. LPA Schedule for All Levels: showing participation of All Levels (from Team Leader to Top Management) with established frequency for all manufacturing areas. 3. LPA schedule is tracked by manufacturing.	• verify if "top level" is conducting the audit. • Check back LPA records, verify that audits were really performed according to schedule and all the operational activities were audited (not only manufacturing operations, but material handling, storage, shipping, *etc.*). • Verify if containment/rework activities are included in the LPA Plan.
A follow-up of the LPA and associated action plans is in place. Deviations are treated	1. LPA records: All the LPA results are documented, including • no deviation found (Y), • deviation found / not corrected during audit (N), • deviation corrected during audit (NC), • not applicable (NA). 2. Problem solving is applied directly at the workstation to treat deviations, actions are defined and recorded with responsibility and a target closing date. 3. Countermeasure Sheet • Deviation found / not corrected during audit (non-conformances or operator claims, as well as safety/ ergonomic issues) must be addressed on Countermeasure Sheet. • Corrective action implementation is followed up. • Countermeasure Sheet is used for Continues Improvement too (*e.g.*, if a good practice is discovered during the LPA, it should be used as a driver to improve the current Standard work). 4. LPA results are used for Continuous Improvement.	• How organization apply the problem solving methodology for issues detected during the LPA and how the organization record the countermeasure (*e.g.*, in the countermeasure sheet) and how they do the follow up of actions: • actions defined against root cause (*e.g.* not only re-training) • due dates kept, • implemented actions were verified. • Check the involvement of the management (knowledge of the result, of the ongoing action plan). Continuous Improvement • Flexibility Chart revised using the results of the LPA • Problem-solving methodology revised using the results of LPA. • Workstation Organization performed. • LPA used to capture a more efficient way to work and lead the standard work revision.

(Continued)

TABLE 9.15 (*Continued*)
An Overview of Criteria for Preparation of an Audit

Required	Criteria of Requirements	Verification Possibilities

LPA effectiveness is continuously monitored and analyzed via LPA results in order to ensure compliance with procedures.	1. Tracking of audit results with visualization to share status on affected area (notice board (nb) of non-conformances per LPA, pareto of nonconformances, repetitive findings, nb of customer issues caused by non-respect of standardized work – it should have been found by LPA).
	2. Tracking of keeping audit schedule and effectiveness of action plan implementation (average time to treat deviations/effectiveness).
		• Perform a layered audit together with team leader. Compare your results with team leader. In case of gap, identify reasons of different evaluation.
		• At shop floor, verify in the visual management board (area or plant):
		  • LPA Plan
		  • LPA Tracking
		  • LPA Results
		  • LPA Action Plan/ Effectiveness

Risk Reduction		
PFMEAs shall be developed and maintained by cross-functional teams for all manufacturing processes and support functions.	1. PFMEA is developed by cross-functional team. 2. PFMEA available for all part numbers and all operations 3. Scoring of Severity/Occurrence/ Detection is according to customer's guideline, ratings are consistent. 4. Content of PFMEA fields is defined properly in accordance with customer guidelines. 5. If supplier is design responsible, DFMEA must be used to develop the PFMEA	• Check member of PFMEA team and verify that necessary trainings are provided. • Evidence that PFMEA was prepared for all the "risky" operations, even though the base was a generic PFMEA. • Scoring is according to predefined standards and customer requirements: • Verify content: Effects evaluated from both customer and manufacturing point of view, Potential cause of failure defined specifically, ambiguous phrases (*e.g.*, operator error or machine malfunction, *etc.*) should not be used. Real preventive actions are listed.
Proactive approach for reduction of PFMEA highest risk items are implemented.	1. For high severity rankings or high risk items, FMEA team ensures that the risk is addressed through existing design controls or recommended actions. 2. Prioritization based on defined approach (*e.g.*, Severity, criticality, and RPN or the combination of Severity, Occurrence, and Detection (Risk Priority Number (RPN)).	• Review risk reduction action plan, evaluate that actions are defined against root cause or improve detection, target dates are kept. • Evaluate whether Quality Tools implemented are efficient in keeping risk under control.

(Continued)

TABLE 9.15 (*Continued*)
An Overview of Criteria for Preparation of an Audit

Required	Criteria of Requirements	Verification Possibilities
	3. Recommended actions documented with responsibility and due date. 4. Quality tools (*e.g.*, LPA, Verification Station, *etc.*) are applied until recommended actions are implemented in order to keep risk under control. 5. Recommended actions re-evaluated after their verification.	• Review scoring after recommended action implemented • Where Severity 9 or 10, detection is low (visual inspection alone is not acceptable). These items must be identified in the Control plan.
PFMEA is reviewed periodically as proactive and reactive activities.	1. PFMEA is reviewed and updated for each past quality issue and corrective action that has been implemented within the target completion date. 2. After each issue, 5 Whys analysis is performed to understand why PFMEA did not predict failure. Action implemented to avoid in future. 3. Periodic PFMEA reviews are scheduled based on process capability, process/product changes, *etc*... which cover: • all processes and their controls are included, • detection ratings are accurate, • occurrence ratings are analyzed using data (SPC, FTQ, ppm, scrap data, Verification Station results, *etc*.).	• Revision date of PFMEA linked to past failures. • Check the last customer complaint or quality issues and their update in PFMEA. • Review checklists, agendas or equivalent that assure adequate PFMEA review. • Check some operations/ processes (material handling, labeling, rework/repair, *etc*...) are included and accurate. • Compare top internal scrap data with Occurrence scoring.
There is a Reverse PFMEA (pro-active approach) process in place to identify new potential failure modes on shop floor.	1. A Cross-functional team conducts on-station review. 2. A schedule of reverse PFMEA (fault tree analysis (FTA) is implemented and regularly updated (timing for review with prioritization of operation and its status / planned-done/). 3. All the findings are driven back into Process Flow, PFMEA, Control Plan, and Work Instructions as applicable.	• Check updates after reverse PFMEA is performed. • Choose one station and perform a quick reverse PFMEA to confirm all current controls are rated properly and all potential failure modes are covered (try to create new ones). • Have the current Occurrence/ Detection numbers been revised?

(Continued)

TABLE 9.15 (*Continued*)
An Overview of Criteria for Preparation of an Audit

Required	Criteria of Requirements	Verification Possibilities
General assessment is conducted to identify potential risks which could impact to plant (normal processes/ activities).	1. There is a formalized process of evaluation and control of the risks concerning resources, facilities, tooling, and supplies. 2. The responsibilities of evaluation and management of risks are clearly established, coherent with the typology of the risks 3. A structured tool to identify, evaluate, and prioritize risk based on a multidisciplinary approach is used (*e.g.*, FMEA approach). 4. Prevention and reaction plans are established and deployed for each identified elementary risk. When appropriate, plans are audited and simulations are carried out. 5. The risks and the associated plans are periodically reviewed based on the plant, corporate, and external lessons learnt.	• Master securing plan which contains major risks, procedures, and owners in case of *e.g.*, flood, fire, *etc*. • Verify that the master security plan covers the relevant major risks. • How risks are evaluated (which criteria, is there a tool like FMEA…). • Who is responsible for the whole process, who is responsible for a precise risk? • An example of risk: Fire. Look at the action plan (preventive: extinguisher, training / corrective: site evacuation plan, sprinkler, firemen on site,...).
Targets are defined and followed to ensure effectiveness of continuous risk reduction activity	1. Tracking the number of high-risk items (trend chart or a Pareto chart). 2. Tracking of delayed recommended actions by leadership necessary actions taken in case of delay. 3. Track that PFMEA reviews and reverse PFMEA are performed based on their schedule. 4. Number of new failure modes and root causes covered after complaint (both external and internal).	• Make sure that the evaluation process is according to your guidelines, as well as the customer's. Start with the highest severity, then criticality, and then the RPN. Try to avoid the RPN analysis as a prioritization of failures – unless dictated by the customer. Avoid RPN numerical thresholds. • Review actions/implementation dates/delays. • Percentage of error proofing/ error detection.

(Continued)

TABLE 9.15 (*Continued*)
An Overview of Criteria for Preparation of an Audit

Required	Criteria of Requirements	Verification Possibilities
Contamination Control		
Sediment is controlled with identification of possible risk areas for contamination, as well as continuous measurement and monitoring of sediment levels.	1. Identify all areas/operations that could be affected by contamination, and failure modes related to contamination are considered in PFMEA 2. Procedure was developed, which includes method defining sediment collection and a system to measure cleanliness. 3. Cleanliness results are documented and plotted on control chart. 4. Control limits are utilized to trigger reaction plans. Do not confuse specification limits with control limits. They are not the same. Control limits are calculated based on the *voice of the process*. Specifications are given by the customer.	• Verify if the organization had mapped all areas/operations that could be affected by contamination and identify the type of contamination in each area. Also, if this information were used during the PFMEA and Process Control Plan (PCP) development. • Check if the cleanliness controls are being done according to frequency established at PCP. Frequencies are reasonable. • Check sediment measurement in lab (instruction and process). • Application using statistical method to monitor and control (Include individual parts, sub-assemblies, and full assemblies). • Verify if there is a documented reaction plan for cases where the limit control or unstable condition is reached. • Ask to team member and/or inspector what they should do when an out-of-control condition is reached.
Sediment reduction strategy is followed to ensure proper functionality of equipment and processes designed to remove/prevent sediment contamination.	1. Procedure/instruction to define and maintain method and frequency of checks to ensure quality and cleanliness of: • Metal Working Fluid System, • Fluid/Air Probes, • Work Station Cleanliness, • Dunnage and Part Storage, • Purchased parts, • Washer, • Debur.	• Check work stations cleanliness, daily verification to be established (Layered Audit, Tape test, *etc.*). • Dunnage cleaning process defined and kept. • Parts are covered / properly protected. • Parts on the assembly lines are handled with lint-free gloves/ towels.

(Continued)

TABLE 9.15 (*Continued*)
An Overview of Criteria for Preparation of an Audit

Required	Criteria of Requirements	Verification Possibilities
		• All wash/debur machine parameters captured and monitored/ documented. • Wash solution and process. parameters monitored and analyzed on an ongoing basis.
If clean room/area is required (due to sediment or painting requirement), special rules are utilized in order to minimize the risk of contamination.	1. Clean room protective clothing is defined and enforced. 2. Positive pressure to stop outside air/ contaminants from being drawn into the clean room. 3. Limited access by employees and enter/exit route to clean room are kept. 4. Sticky mats or shoes protection to remove contaminants from footwear. 5. Control of chemicals detrimental to the process.	• Check if protective cloths are worn (*e.g.,* hair nets, shoe covers, lint-free lab coats, rags, gloves, *etc.*) • Check how access control to clean room is kept. • See if air quality and pressure are monitored. • Verify if filters are changed according to planning. • Check if controls of restrictions (*e.g.*, no fibrous material, no aerosol sprays, no food, *etc.*) are in place. • See if LPA is in place (if out of out-of-condition is identified, there's a reaction plan in place).
Maintaining the processes to control and prevent dirt in paint or parts and foreign material contamination.	1. Sources of dirt and foreign material as a potential failure mode are considered in PFMEA. 2. Methods developed to prevent extra parts or materials that may fall into or stick to products and verified through error proof verification and/ or layered process audits (*e.g.,* sensor for masking used). 3. Procedure or instruction describes dirt control and prevention for key areas: People, Process, Facility, and Material. 4. Application of statistical method to monitor and control the painting process, such as dirt count (SPC, U-chart), dirt identification (Pareto), *etc.*	• Check if foreign material/extra part (screw, pin, washer, any small components, *etc.*) effects (like noise, function issue, etc.) are considered in the PFMEA. • Verify that dunnage, fixtures, conveyors, washers, and pre-assemblies are protected to prevent falling material. • Check possibility of assembling the part without masking. • Verify if all stations in the paint area are clean up and operators do a self-inspection of the area. • Ask a team member to explain what are the main sources of dirt – key issues in the people, process, facility and material area and how he works to avoid them.

(Continued)

TABLE 9.15 (*Continued*)
An Overview of Criteria for Preparation of an Audit

Required	Criteria of Requirements	Verification Possibilities
	5. Ensure proper functionality of equipment and processes designed to remove/prevent dirt in paint or parts. 6. Behavior Instructions are defined and respected.	• See if the sources related to process (paint spatter, overspray, towels, etc.) are checked by PM, and there's a reaction plan established. • Check if general housekeeping process is established for facility and is followed.
Targets are defined and followed to ensure continuous contamination reduction activity	1. Sediment result monitored, reduction plan defined. 2. Layered audit and process-specific audits results. 3. Internal or customer issues related to cleanliness. 4. Pareto of defects for key sensible process (painting, metallization, etc).	• Review sediment results and verify sediment reduction activities. • Evidence of site leadership reviews. • Audits performed focusing contamination control.
Supply Chain Management. *(The intent here is to make sure that ALL materials have been ordered with appropriate specifications and are available for use).*		
Procedures or work instructions to select and evaluate suppliers are implemented.	1. Procedures that require supplier audits to support sourcing decisions, evaluation of their systems. 2. Criteria to become a sourceable supplier are defined. 3. Approved supplier list to track tier suppliers' performance (*e.g.*, Bid list). Performance indicators are defined with threshold. 4. Contractual documents (such as system of record – SOR) contain all tier1 requirements, expectations, and processes, signed by tier supplier. 5. Supplier Quality Department or organization is fully involved in the sourcing process.	• If supplier selection is done at manufacturing location, check documentation of the tier supplier selection process. • If selection is done centrally, check tier X status is known by manufacturing location. • Evidence of SQ responsibility (SQ organization chart or SQ training records). • The effect of manufacturing location on tier suppliers' performance evaluation.
Customer-specific requirements are cascaded to all tier supply chains and deliverables of tier X are validated/ approved by tier 1.	1. Customer-specific requirements are implemented at tier suppliers: • technical specification of product, • product and process specific standards, • procedures need to be applied (*e.g.*, PPAP/EI, PFMEA...), • traceability, FIFO, and labeling requirements.	• An example of request for quotation. • Technical specifications forwarded to tier X, including list of critical characteristic in their PCP • Supplier quotation file, including technical/process feasibility

(Continued)

TABLE 9.15 (*Continued*)
An Overview of Criteria for Preparation of an Audit

Required	Criteria of Requirements	Verification Possibilities
	2. Tiered suppliers comply with product and process validation requirements. Approval is available for all the components needed for assembled part. 3. Final product key and pass-through characteristics are identified on component (or material) level as well as shared and controlled (if required with SPC) with tier supplier. 4. All the design and process changes are communicated, managing change requirements are applied. Tier X supplier production transfer needs to be managed with specific procedures for all their suppliers. 5. Tier suppliers' capacity is confirmed for all the materials/components needed for assembled part.	• Check critical components product/process qualification files (PPAP). • Check customer drawing whether it contains any key/pass through characteristics, which are produced and controlled by tier X supplier. If yes, verify that tier X supplier drawing, PPAP, Control Plan cover them, as well as tier 1 incoming control includes them. • Example of management on a tier X process change. • Check capacity confirmation for critical components.
An escalation process is applied in case of tier supplier issues	1. For any delivery of parts with deviation, tier 1 validates the formal deviation request of the tier X complaint. 2. For single issues, a systematic and disciplined approach to problem solving is implemented, an action plan is set up by tier X and verified by tier 1, keep due dates expected by customer. 3. Approved supplier list to track tier suppliers' performance (*e.g.*, Bid list six panel). Performance indicators are defined with threshold. 4. In the event of non-respect of a target, an action plan to reach it is systematically defined and followed by tier X supplier. 5. When a tier supplier moved from green ranking, escalation process takes into account the tier 1 support of the tier X supplier to improve. Exit criteria are defined.	• Prior to audit check whether there was any customer complaint caused by tier supplier. • Example of supplier complaints toward its tier 2, verify contents of the complaints • Is there a timely management with milestones? • Verify content of 5 Whys prepared for customer complaint with problem resolution report (8D; 5P; Red-x, *etc.*) • Example of action plans given by tier X. How is it validated and verified? • Example of supplier audit, evidence of audits at key suppliers • Escalation criteria (*e.g.*, Controlled Shipping, Top Focus, New Business Hold, *etc.*)

(Continued)

TABLE 9.15 (*Continued*)
An Overview of Criteria for Preparation of an Audit

Required	Criteria of Requirements	Verification Possibilities
	6. Regular audits are performed at supplier's plants. In case of complaint, to verify action implemented, and regular audits at key suppliers to improve their system (CQI audits, process-specific audits, PCPA...).	• Critical suppliers identified and tracked via processes such as the Top Focus process, exit criteria defined. • Review tier X performance matrix with tier X corrective action plan approved by tier 1.
Incoming inspection process in place. (A preferred plan would be the use of SPC data.)	1. An incoming inspection control plan and the related control standards and sampling rules are established for each key/critical supplies' product. 2. Only approved (PPAP/EI and incoming inspection released) components/materials are used for assembled part. 3. At the reception of supplied product, the first controls take into account the checking of the quantity, the integrity of the packaging, and the identification of the product. 4. Part inspection requirement, including Pass Through, is established and implemented for each supply since the design phase and updated in case of non-conformity.	• Check identification of material released for production after receiving. • Verify that incoming covers key and Pass-Through characteristics, frequencies are reasonable. • See a nonconformance found during incoming. • Verify link between incoming frequency and supplier performance (high ppm supplier). • Check incoming records for a component.
Tier supplier targets are defined and their performance are tracked via indicators	1. Tier X performance result (ppm, rate of complaints, delivery, *etc*). 2. Mean time to respond to complaints. 3. PPAP curve (planned full PPAP date vs. achieved). 4. Metrics on bid list (number of red suppliers, *etc.*). 5. Tracking results of the audit result at tier X. 6. Average time to close Problem Resolution.	• Check customer complaints, ratio of tier X supplier issues. • Red suppliers on the Bid list, actions for improvement. • If any significant concerns about tier X issues had been addressed in the FR meeting.

(Continued)

TABLE 9.15 (*Continued*)
An Overview of Criteria for Preparation of an Audit

Required	Criteria of Requirements	Verification Possibilities
Managing Change		

All product, process, or source changes are monitored and controlled	1. A procedure to manage product/ process /source changes is defined and applied – both planned and emergency changes.
	2. A change form is utilized to document all changes and controlled through a Document Control Process (*e.g.*, tracking log sheet, revision numbering system, approval process, *etc.*).
	3. All changes are managed like a project; responsibilities and milestones are defined, planning, activities, and the deliverables are established in agreement with the customer.
	4. All changes need to be reviewed and approved by customer (PPAP and or engineering). Inform customer in the early phase to get approval before kickoff.
		• Check documentation via example: A design change and a process change:
		  • organization (project team) and milestones,
		  • evidence of customer approvals,
		  • planning and evidence of reviews,
		  • tool to ensure traceability of modifications.
		• Evolution of data systems, such as MRP systems, storage management software, and EDI server must be considered as major changes.

A risk analysis is applied for any product/process change.	1. For any product/process/source change, a feasibility analysis is carried out. The study takes into account the impacts in terms of costs, technical, performance, quality, timing, and capacity. 2. Potential risks of change are considered via FMEA approach and documented in DFMEA and PFMEA. 3. Based on risk analysis, planning and implementation of the change are carried out, and a product and process validation plan is defined. 4. Break point is defined when change become irreversible, it is communicated to customer. 5. According to risk analysis, a securing approach is put to work in the launch phase (*e.g.,* mixing of old and new design).	Check one of last of modification and verify: • A feasibility analysis, including lead time analysis • Formalized impact evaluation and risk management • PFMEA and DFMEA (if applicable) • Action in place to cover risk identified

(Continued)

TABLE 9.15 (*Continued*)
An Overview of Criteria for Preparation of an Audit

Required	Criteria of Requirements	Verification Possibilities
A Production Trial Run (PTR) process is implemented.	1. A standardized communication procedure and form to control and monitor all Production Trial Runs (PTR) that documents each step of the process and records all approvals and results. 2. Traceability of trial run batch is ensured. 3. Quality reviews documented to release product for PTR shipment, and verification process has returned to normal production.	• Example of a trial run • Trial run validation criteria and results (records)
Maintenance		
Maintenance organization and strategy are established and deployed.	1. Maintenance process is formalized and covers all the machines, tools, devices, equipment, and facilities on the site. It includes preventive and corrective maintenance. 2. A system for managing, planning, organizing, and monitoring maintenance operations is set. 3. Resources are available during all production periods and outside production periods to ensure non-critical maintenance operations (*e.g.*, preventive maintenance). 4. Suitable maintenance facilities and equipment are available. 5. Fast communication between production and maintenance is assured. 6. A process is in place to improve production output. It is based on analysis of operational availabilities, operators' suggestions, 6-sigma techniques, *etc*. 7. Approach to standardize equipment (*e.g.*, using the same filter, the same interfaces to tools, *etc*.).	• Preventive maintenance for safety equipment has to be performed on time without exception. • Check that a maintenance system in place to manage all maintenance activity, supported by IT tools like CMMS (computer maintenance management system), Excel, and so on. • Available resources by technology, including outsourced experts (flexibility chart). • Implementation of resources near the manufacturing activities. • Are there free resources to manage the corrective maintenance? • Facilities available (areas well defined, conditions, 6S level, *etc*.) • Management improvement strategy and periodical reviews. • Review major brand types of major equipment – ask about process to purchase new machines. • A generic plan to improvement of a type of equipment (*e.g.*, electrical screw drivers).

(Continued)

TABLE 9.15 (*Continued*)
An Overview of Criteria for Preparation of an Audit

Required	Criteria of Requirements	Verification Possibilities

The activities of maintenance are planned, performed, and tracked.	1. The Preventive Maintenance planning takes into account risk classification of the equipment (safety, constraint, bottleneck equipment, unique process with no substitution process, *etc*.). It is built together with the Master Production Schedule (MPS), which takes into account time allocation for maintenance.
	2. The planning is regularly followed, updated (including outsourcing maintenance).
	3. Corrective maintenance is carried out for any deviation from the nominal (according to manufacturing priorities). The Preventive Maintenance activity is periodically reviewed based on the corrective maintenance activities.
	4. Maintenance activities are considered as Standardized Work. Technical documentation is available and managed for all equipment.
	5. Records of all maintenance activity and results are formalized and filled in.
	6. Necessary re-qualification is done after maintenance activity, and its result is recorded to ensure traceability.
		• Choose machines (constraint/complex one) and a tool to verify:
		• Maintenance planning: Identification of the equipment, task to do, when
		  • Follow up of the maintenance schedule and its visual management
		  • maintenance work instruction (including changes due to lesson learned)
		  • technical documentation for a precise equipment including document management
		  • records of corrective maintenance activity (type of equipment, is there repetitive breakdown,...)
		  • records of preventive maintenance on an equipment with a recent breakdown
		  • Verify that maintenance activities are fully deployed and cover all equipment (machines, facilities, tools)
		  • Different types of preventive maintenance depending on the type of equipment (pure preventive approach and conditional maintenance).

Level 1 maintenance is systematically applied and integrated into workstation instructions.	1. Level 1 (L1) maintenance operations take into account equipment identification, cleaning, self-maintenance, and safety devices verification. 2. L1 maintenance is performed under manufacturing responsibility at operator's workstation. L1 operations are integrated in the workstation's work instructions.	• Ask operators about Level 1 maintenance responsibilities. • Verify few L1 maintenance working instruction and record. • Look at the equipment status on the shop floor (is there an identification (number), cleanliness, protection in good condition, leakage, …).

(Continued)

TABLE 9.15 (*Continued*)
An Overview of Criteria for Preparation of an Audit

Required	Criteria of Requirements	Verification Possibilities
	3. Any deviation, anomaly, or improvement suggestion is recorded and, if necessary, escalated to a higher-level maintenance activity. 4. Records are analyzed and used as lessons learned to improve maintenance operations.	• LPA records. • Evidences of activities transfer from preventive maintenance to L1. (The levels of maintenance are: L1- unit; L2- direct support; L3- general support; L4- Depot).
The spare parts and their storage are managed. The critical parts are identified.	1. A list of critical spare parts is determined, managed, and regularly updated. 2. A spare parts stock is available with a minimum stock level. Spare parts tracking system is combined with maintenance system in order to control physical inventory. 3. The spare parts are stored in suitable conditions, and periodic physical inspections are performed for long-term stored items.	• List of critical spare parts. • Stock of spare parts: reception / organization / consumption • Conditions of storage. • Computer-aided system. • Inventory (take an example of a critical part and verify the robustness of the inventory).
The customer's product-specific tools are managed to preserve tool condition till the end of their lifetime.	1. Each Customer Specific Tool (CST), including customer-owned one, is identified in a single and inalterable way according to customer requirements (customer marking). Identification includes tool change level. 2. Lifetime of tools is strictly followed and documented in "diary sheet," which includes lifetime of tool, shoot number, maintenance activities. 3. In case of subcontracting, a list with the localization of the CST is established and communicated to the customer. 4. The storage of the CST is organized and managed. Activities to be carried out prior to storage are defined. Storage conditions guarantee the safeguarding of the CST until its end-of-life. 5. Tool release process after storage is established. 6. Product/process re-qualification is carried out for each replacement tool according to customer requirements.	Check few tools to verify: • General conditions of the tools (leakage, rust, ...) • Identification and visual management of the tools • A diary sheet and records, lifetime followed • Communication and customer approval for each change on an example • Storage conditions • Standardized maintenance operations

(Continued)

TABLE 9.15 (*Continued*)
An Overview of Criteria for Preparation of an Audit

Required	Criteria of Requirements	Verification Possibilities
Indicators are defined and tracked to ensure effectiveness of all the maintenance activities.	1. Performance and reliability targets are defined and based on historical data and are related to indicators, which are tracked • Failure rate, MTBF, MTTR, stop of lines. 2. Tracking maintenance performed vs. planning (including service provider activity). 3. Pareto analysis of breakdown. 4. Average rotation of spare parts. 5. Deviations found during spare parts inventory audit. 6. Ratio of corrective maintenance against preventive maintenance.	• Customer complaints caused by machine or tool problem (*e.g.*, burrs issue…). • Verify quantity of a few spare parts in stock. • Check a few maintenance working instructions whether they are standardized.
Manufacturing and Material Flow Management		
A structured process of manufacturing scheduling is implemented and systematically reviewed	1. Long-term strategic scheduling is managed via a Sales and Operating Planning (S&OP), which includes a complete forecasting of customer demand. 2. The S&OP is reviewed regularly and is used to define manufacturing capacity, stock level, and investment plans. 3. A master Production Schedule (MPS), coherent with the S&OP outputs, is managed on the site. It provides a complete forecasting of the customer demand at the lowest level (part number) in short term. 4. The MPS is reviewed regularly by logistics and manufacturing team (supported by maintenance) and is used to define resource needs (equipment time allocation, human resources, material). 5. S&OP and MPS are transferred to concerned entities on the site and to tier X.	• Check long (SOP) and short master production schedule (MPS) term scheduling for a product. Verify that: • capacity meets long-term customer demand, • time allocated to other product is part of review, • SOP regularly reviewed based on electronic data interchange (EDI) data, • long-term demand deployed into short-term scheduling, • check a tier X demand whether it is in line with tier 1 manufacturing schedule. • Supplier should have at least a sheet with the list of constraints in the plant and up-to-date availability information (opening time, preventive maintenance planning).

(Continued)

TABLE 9.15 (*Continued*)
An Overview of Criteria for Preparation of an Audit

Required	Criteria of Requirements	Verification Possibilities
The daily manufacturing activity is planned in detail and followed up at site level.	1. A process is in place led by manufacturing to generate manufacturing detailed program on daily basis, coherent with MPS outputs. 2. The program takes into account the preparation time batch change, tool change, set-up, start-up process, ...) and up-to-date OEE Rates. 3. Deviations between forecast and real production are followed and controlled on daily basis at production line level. An escalation process is defined. 4. A process to improve the setup time is in place. Organization should establish a goal, measure the setup time, and define the action plan once the setup time goal is not reached.	• Check product chosen in production section and check daily deployment: • manufacturing detailed program, • complete preparation time for a batch, • different programs from a week to another (is it stable or not, level of flexibility). • The real production level. • Planning and minutes of optimization workshops. • Hourly Board Control. • A strategy is in place to improve the setup time (*e.g.,* SMED strategy).
Constrained operation is identified and specifically managed.	1. Based on customer demand all the bottleneck operations are identified. 2. Constraint operation identified among bottlenecks and prioritized in regard to qualified operators, training, maintenance, scrap, setup, and fast reaction in case of any deviation. 3. Manufacturing detailed program is optimized, taking into account constrained operation. 4. Constraints shall be managed by identifying problems, establishing action plans, and verifying effectiveness of action plans on a regular basis. 5. A backup plan for each bottleneck operation is defined.	• List of bottleneck equipment. • Check via record that: • constraint running continuously, • number of qualified people via Flexibility Chart • breakdowns and actions against them. • Check buffer size, ask operators how often they run out.

(Continued)

TABLE 9.15 (*Continued*)
An Overview of Criteria for Preparation of an Audit

Required	Criteria of Requirements	Verification Possibilities
Product packaging (final product, intermediate, and supplies) availability and conditions are managed to ensure product quality.	1. Loops or flows of full/empty packaging are organized and managed. 2. The stocks of empty packaging are organized and managed. The number of empty packaging available is enough to keep the production running. 3. A process to check the packaging is applied; it takes into account the requirements of the customer and the standard definition of the packaging. All activities are considered as Standardized Work (SW). 4. Packaging control work instructions are established and applied, which include decision criteria for the use of packaging, the associated countermeasures, and corrective actions (*e.g.,* alternative packaging, cleaning, maintaining, special transportation of empty packaging, *etc.*). 5. Supplier Responsibilities are clearly defined. Deviations are systematically communicated and treated with the customer.	• Examine the flow of empty packaging. Try to see the loops are respected. • Verify the procedure to initiate the supply of additional packaging (threshold definition, alert process, timely response). • Check empty packaging storage area. • Verify a work instructions for checking and cleaning. • What happens if there are no more empty packaging? • Look at the conditions of the packaging particularly the label support. • Check actions for critical packaging from availability point of view (*e.g.*, Alternative packaging is defined, for final product it is approved by customer).
Handling and storage conditions of product (final, intermediate, and supplies) are respected in order to protect parts from damage and environmental affects.	1. All storage areas are identified and organized by type of product (incoming, intermediate, Work In Process, final, *etc.*). Process Flow charts effectively match processes. 2. Potential failure modes related to material storage and handling are considered in PFMEA (*e.g.,* damage by handling equipment, rust caused by storage conditions, mixing up of products, *etc.*)	• Check storage area and condition at several places (incoming, Work in Process, final product). • Ask operators and material handling personnel whether they are aware of and following instructions. • Verify that adequate protection in place to protect parts from damage and mixing. • Condition of storage (temperature, waterproof,…).

(Continued)

TABLE 9.15 (*Continued*)
An Overview of Criteria for Preparation of an Audit

Required	Criteria of Requirements	Verification Possibilities
	3. When needed, storage conditions are controlled by devices in real time (*e.g.*, temperature, humidity, *etc.*). Records of stock condition are kept. Alert procedures and countermeasures are defined. 4. Standardized Work (SW) is applied, structured approach for organization of storage are defined and applied. System in place allows to visualize easily storage operations and level of stock for each reference. 5. Associated equipment is suitable for stocking and handling. 6. The stock management system takes into account: Product expiring dates, product change level, and the respect of the FIFO. 7. Stock level is checked via regular quality and quantity audits (*e.g.*, LPA).	• Visual management in place (level mini and maxi are visible, token board for the filo). • Check expiring dates by reading labels. • FIFO: risky situation to examine: intermediate stocks, double flows/lines (*e.g.*, 2 paint lines): how do they manage this type of situation (specific rules and procedures). • Specific management for the high runner references. • Results of audit or inventory. • Equipment to manage the stock (barcode reader, informatics systems.
A system is in place which ensures that materials needed for production are organized and available at place of their application.	1. Feeding plan is established and tracked. Suitable feeding equipment adapted to layout, to flow configuration, and to available space are used 2. The feeding process is based on a structured methodology (*e.g.* pull system) which: • minimizes overflows & non value-added operations (repacking) • guarantees the availability of materials at the workstation during all • production period 3. All the feeding activities are considered as Standardized Work (SW) and guarantee the respect of FIFO. 4. At the workstation, materials are used in manageable sizes and material flows are organized and managed following the standard work/in-stock process.	• Check if a standardize feed route plan is defined and followed. • Check few workstations and check: • Respect of the "pull system" principles, • Alerts from the lines and their management: Management of the risk of stockout?, • Work instructions (line feeding operations, loading, repackaging operations…), • FIFO kept (organization of the rack). • Organization of "supermarket" areas (visual management) if existing. • Organization of the material flow at the workstation (entry and exit points for each component, packaging, and useless movement).

(Continued)

TABLE 9.15 (*Continued*)
An Overview of Criteria for Preparation of an Audit

Required	Criteria of Requirements	Verification Possibilities
	5. Material flow optimization is in place based on the use of tools such as MIFA or VSM.	• Optimization activities (minutes of meetings, action plans...).
Indicators are defined and tracked to ensure effectiveness of material flow management.	1. Indicators concerning the availability of packaging. 2. Stock management indicators: • Inventory, • Stock targets fulfillment. 3. Lines feed indicators: • Feeding tracking, • Internal service rate, • Production losses due to feed missing, • Internal flows lead time. 4. Tracking of constrained output.	• Prior to audit, check any customer complaints issued related to packaging and material handling. • Layout of areas that are optimized. • Where applicable, the edges of lines are fed automatically or by little trains (no big boxes or full of components). • Stock level (min, max visual mgmt). • Workplace visual mgmt of stock (min–max).
External Logistics		
Supply process (incoming) is managed, organized and tracked.	1. A process to follow deliveries is formalized and applied. Related operations are standardized. Resources are available and properly managed. 2. A master schedule of deliveries is defined and follow up. A visual management system, showing the smoothing of the deliveries (leveling), allow to follow them. An escalation process is implemented in case of deviation. 3. A logistic protocol is defined and regularly updated for each tier X supplier. Transportation organization toward the tier X suppliers is optimized. 4. The service rate of the tier X suppliers is followed; actions plans are established for most failing tier X.	• Organization of the incoming. • Master schedule of deliveries (stability through weeks). • Working instructions. • An example of protocol with a tier X. • Service rate metrics. • An example of alert in case of deviation. • Resources available matches incoming and outbound deliveries.

(Continued)

TABLE 9.15 (*Continued*)
An Overview of Criteria for Preparation of an Audit

Required	Criteria of Requirements	Verification Possibilities
A process to secure supplies is applied on a basis of risks analysis.	1. Components/materials at risk of supply are identified. Securing plans including safety stocks are established with the tier X. 2. Securing plans are periodically reviewed based on lessons learned. 3. Safety stock, advance warehouses are contractually defined and managed for the "distant" tier X (more than 48 h of transportation) and when applicable for the "risky" tier X. 4. The management of the site is strongly involved and validates the whole process.	• Identify a "risky" supply (*e.g.*, distant supplier or failing supplier). • Examine the securing plan associated. • Emergency procedures with alternate supply process. • In case of safety stocks, identify clues of contractual definitions.

Shipping process (outgoing) is organized and tracked. Packaging is properly labeled.	1. Shipping process is described and timely managed. All the operations are standardized.
	2. Shipping process integrates milestone operations (*e.g.*, verification of availability of finished products, truck loading completed (*i.e.*, Lear corporation's system AVIEXP which has a purpose to provide the final consignee of the goods, with detailed information relating to the actual contents of a consignment sent by a consignor)) allowing to alert customer in relevant time in case of issue.
	3. Shipping planning is visually managed (Timing Table of the preparations of shipping orders).
	4. A preparation list is available for operators. Operations done are recorded. Loading is prepared on "bogus truck" areas.
	5. All the packaging is labeled properly, and packaging/shipping label error proofing operation is a best practice in order to prevent part misidentification. PFMEA covers potential failure modes of labeling.
		• Shipping management board.
		• Verify that the shipping process includes at least the following milestone:
		  • verification of finished products availability,
		  • start of preparation,
		  • end of preparation. Ready to ship,
		  • truck loading completed (EDI message sent to customer).
		• Identification of deviations on the board (late supply).
		• Organization of bogus truck areas.
		• Preparation lists.
		• Labels correctly fulfilled (with the right routing code).
		• Equipment and work instructions to check the product and pallet labeling.
		• Hooping of the pallets.
		• Emission of the EDI message AVIEXP.

(Continued)

TABLE 9.15 (*Continued*)
An Overview of Criteria for Preparation of an Audit

Required	Criteria of Requirements	Verification Possibilities
	6. Specifically for synchronous deliveries, a constant monitoring/ follow-up of the process is carried out in real time; all deviations/issues are recorded and analyzed, and immediate countermeasures and corrective actions are started to eradicate issues.	
In order to guarantee that shipping process is carried out according to customer needs, the Electronic Data Interchange (EDI) network is operational and a Logistic Protocol is managed.	1. EDI communication is installed and is validated with the customer. Qualified people are permanently available on the site to manage EDI. 2. Back-up solutions are defined, validated with the customer and are periodically tested. 3. Any evolution of EDI communications must be considered as significant change. 4. A Logistic protocol is established together with the carrier (transporter) and the customer. It is continuously updated to take into account changes during current production (change of customer site, change of schedule, *etc.*). Resources responsible for the protocol management on-site are identified. 5. Any deviation in the protocol application is treated and managed by action plan (alternative transportation mode, adapting packaging, *etc.*).	• Several logistical electronic protocols (LEPs) and validation status. • Coherence between different LEPs and shipping schedules. • EDI equipment. • Securing equipment (hardware, backup, networks). • What happens if EDI network falls? • Customer site contact list.
Indicators are defined and tracked to ensure effectiveness of external logistic management.	1. Indicators concerning the supply process: • Fill rate of the trucks, • Service rate of tier X Suppliers, • Tracking of Logistic issues with tier X Suppliers. 2. Indicators concerning the shipping process: • Shipping lead time, • Customer service rate, • Tracking of customer Log issues,	• Prior to audit, check any customer complaints issued related to reception issues, stockout. • visual management (indicators are followed on boards at the shop floor). • How supplier determines safety stock level – risk analysis?

(Continued)

TABLE 9.15 (*Continued*)
An Overview of Criteria for Preparation of an Audit

Required	Criteria of Requirements	Verification Possibilities
	• Pareto of failures (customer line stops / stockout), • Rate of mislabeling. 3. Indicators concerning the safety stock (applicable only if a safety stock has been contractually defined with customer) • Stock level, • Rotation index of the safety stock.	
Strategy and Project Management		
The industrial strategy vision and the associated roadmap are defined and deployed.	1. There exists an industrial long-term vision (2–5 years) coherent and communicated to the whole staff. 2. Long-term vision is deployed to strategic targets of the site. 3. A robust process to transfer targets until the operational level (from the site targets to the operator targets). 4. Based on target “Improvement Master Plan” developed, which includes the modification of the internal processes and takes into account the organization, the resources, technology, the evolution of product/process, the evolution of skills, *etc*. 5. The Improvement Master Plan is managed by Site Manager Director. Actions are followed by the site top management team and, when applicable, by corporate industrial central department.	• Check long-term vision for site and strategic targets deployment for few departments. • Evidence that performance is reviewed and action is initiated in case of deviation.
The organization of the manufacturing site is completely described for all key activities.	1. The manufacturing site is involved in new projects since advanced stages, at least project team includes staff from manufacturing site. 2. A project milestone states the transfer of responsibilities between project and production site. After transfer to manufacturing location process needs to be re-qualified.	• Check site management org chart. • Check project responsibility, site contact part of it. • Check few job descriptions, one of them for quality manager.

(Continued)

TABLE 9.15 (*Continued*)
An Overview of Criteria for Preparation of an Audit

Required	Criteria of Requirements	Verification Possibilities
	3. An organization chart details the site organization and is systematically updated on the basis of changes in organization and missions. 4. A job description is available for each job and systematically updated on the basis of changes in organization and missions. 5. Quality is independent of the operational directions. It has authority to stop the production in the event of nonconformity.	
Human resources for all the operational activities are evaluated and continuously managed.	1. There is a formalized process to identify new resources needed. 2. Rules are defined for their implementation in the event of modification of organization or process, evolution of production program, or new product launch, *etc.* 3. The process integrates management of the rates of new employees, including ratio of the temporary staff based on indicators of production schedule and efficiency.	• Check whether new project requires new staff. • Check process for identification of needs (number of employees, their qualification). • Pre-defined plan based of production need (new shift/ additional operators, *etc.*).
For all new facility/ equipment/tools, *etc.* (manufacturing, control, logistics, *etc.*) or in case of revamping, a structured validation approach is applied.	1. Milestones are established and aligned with project timing for installation and validation of new equipment/tools, *etc.* 2. Based on milestones detailed timing plan is established and regularly followed up for validation process. In case of any delay/deviation, actions initiated, including notification of customer. 3. The qualification process aims to verify compliance with specifications, contract, and performance (productivity – capability – quality) of equipment. In the case of gages, it also integrates equipment calibration. 4. Dedicated resources assigned for qualification of equipment, connected to maintenance.	• Check whether new project requires new facility/equipment/ tool, *etc.* • Check process for identification of needs. • Check new equipment and its documentation for qualification, technical data, and maintenance.

(Continued)

TABLE 9.15 (*Continued*)
An Overview of Criteria for Preparation of an Audit

Required	Criteria of Requirements	Verification Possibilities
For every new product/process or significant product/process changes, the product/ process qualification is defined, continuously tracked, and finally approved.	1. Milestones are established and aligned with customer project timing for product/process qualification. 2. Based on milestones detailed timing plan is established and regularly followed up, their results are shared with customer. 3. Qualification process includes all manufacturing processes (including outsourced), components, tier parts, and raw materials. 4. Deliverables to the customer are managed and controlled. Customer Product Approval Status is known and tracked for all the products. 5. Capacity analysis with its verification, process audit, and capability of key characteristics are part of the qualification process. 6. Any deviation is treated; actions plans are led by the project manager and / or his representative on site. Customer is notified.	• Check a timing plan for an advanced project. • Check a qualification package (PPAP) and cross-check few elements (*e.g.,* gage R&R, dimension and material report, function/durability result, capability study). • Check a R@R result. • Check qualification (PPAP) confirmation from customer.
Service parts (SP) production process is managed and controlled during and after the mass production phase.	1. A process to manage "end-of-life" of current production of reference part number is deployed. It takes into account the quantities of SP remaining to produce. (strategy defined, *e.g.*, prepare stock to cover SP demand, keep tool, production transfer, *etc.*). 2. In order to maintain efficiency and effectiveness of production and measurement equipment dedicated to SP, specific procedures of storage and maintenance are applied (*e.g.*, storage conditions, 6S, oiling, protection against dust,…)	• Business planning + production master plan (verify that SP needs are included). • Example of "end of life" production. • Process qualification of specific operations. • SP storage. • Procedure for long storage.

(Continued)

TABLE 9.15 (*Continued*)
An Overview of Criteria for Preparation of an Audit

Required	Criteria of Requirements	Verification Possibilities
	3. Specific operations or processes for SP (*e.g.*, specific marking) production are industrialized, managed, and controlled according to the same methods as mass production process. 4. In the event of storage of SP for a long period before shipping, a process is applied in order to guarantee the safeguarding of the parts (intermediate packaging, new storage areas, *etc.*).	
Product packaging (final product, intermediate, and supplies) and handling equipment are designed, validated to guarantee product conformity.	1. Packaging is defined during product design phase (take into account in project milestones). The technical specification is built on the basis of customer needs and is approved by the customer (drawing of packaging). 2. Packaging is designed in order to protect product from damage. 3. Test procedures and validation process are shared with the customer. Static and dynamic tests are implemented. Test reports are available. 4. The definition of the storage and handling conditions starts during product/process design phase and is shared with the customer. 5. Handling and transportation equipment design is done taking into account risks of degradation of product quality.	• An example of technical specifications. • An example of packaging. • Packaging part protection material. • PFMEA concerning handling and storage. • Storage control plan when necessary. • Test result (static & dynamic).
Part and Process Control		
There is a system to guarantee the availability to users of the documents at the latest version at the right place. All the technical information is available.	1. A standardized process guarantees that all the users have access to the latest versions of the documents. 2. Revision history available for each document. 3. Customer-approved drawings are available at production facility with the latest change level.	• Prior audit check drawing date and engineering change level of the latest customer drawing. • All technical regulations referred to in drawing are available.

(Continued)

TABLE 9.15 (*Continued*)
An Overview of Criteria for Preparation of an Audit

Required	Criteria of Requirements	Verification Possibilities
	4. Drawing is completed with all the product characteristics (tolerances, GD&T, Correct datums, key process characteristics (KPCs), *etc*.). 5. All technical regulation are available at production facility.	
All the key characteristics are identified in order to ensure their proper control.	1. All the key characteristics defined by customer are identified in PFMEA and Control Plan (CP). 2. Key customer characteristics are deployed to process characteristics if applicable. 3. If applicable, they are communicated to tier suppliers, necessary controls are verified and confirmed. 4. Required customer controls (*e.g.*, SPC) applied for key characteristics. 5. Pass characteristics are identified and controlled. 6. Error proofing strategy is applied.	• Check all the key characteristics deployment into documentation (PFMEA, CP, and working instructions), verify their control effectiveness. • Ask operators about awareness of key characteristics.
All the operations are listed and charted in Process Flow diagram.	1. All process steps, including logistic flow, are listed and charted. 2. The flow chart is used as an input for PFMEA. 3. Link ensured between Process Flow, PFMEA and Control Plan.	• Process Flow includes: receiving, all the operation steps, rework, scrap, gauging/inspection, shipping, labeling and Part ID at receiving, waste in process (WIP), finished good and shipping • Flow is clearly arranged. • Workplace is properly configured and matches with Process Flow Diagram. • Check link/number matching. between Process Flow, PFMEA, and Control Plan.
Control Plan is available for each product in order to define necessary controls need to be performed.	1. A control plan is available for each product and process. 2. All Current Controls are listed on the PFMEA detailed in the Control Plan.	• Control plan addresses print requirements relative to (dimensional, material, rust preventative, heat-treat, micro-finish, contamination/ sediment, *etc*.). • Sample sizes and frequency for each operation are reasonable (*e.g*., based on internal data: scrap, rework, FTQ, *etc*.).

(Continued)

TABLE 9.15 (*Continued*)
An Overview of Criteria for Preparation of an Audit

Required	Criteria of Requirements	Verification Possibilities
	3. Control plan defines the characteristics and the parameters to be monitored, the specifications and the tolerances, the control frequencies, the methods, the equipment, the records, and the reaction plans (in the event of nonconformance). 4. Sample size and frequency take into account process and measurement capabilities, phase of project, and how often reaction is needed. 5. Control plan is reviewed systematically in the event of PFMEA review.	• Evidence that Control Plan is being updated. • Checks to be done under the PCP requirements are performed.
Statistical Process Control is applied.	1. Characteristics, where SPC needs to be applied, are defined in Control Plan based on customer and process capability requirements. 2. Sample size and frequency are defined according to the stability of the process. 3. Control limits are calculated for each characteristic according to the capability target and updated based on trends. 4. Results are documented via control charts and used for process capability calculation. Result needs to meet customer's requirements. 5. Out-of-control point is noted with corrective action taken. 6. Leadership periodically reviews SPC results and ensures that out-of-control conditions are managed.	• Where the Process Control Plan (PCP) calls for SPC look at actual SPC records and data collection: • all data boxes filled in • control limits calculated • there trends/patterns present that would indicate a need to recalculate limits • the limits actually are control limits, not spec limits. • Verify that every out-of-control points are noted and corrective action defined and followed by leadership. • Ask someone noted on the action plan to explain their involvement and or responsibility. • Manufacturing process demonstrates the required capability or performance (based on SOR requirements).

(Continued)

TABLE 9.15 (*Continued*)
An Overview of Criteria for Preparation of an Audit

Required	Criteria of Requirements	Verification Possibilities
Results of control are recorded and archived, as well as analyzed to correct deviations and improve processes.	1. Control records are documented (using standard forms or Information systems) according to Control Plan requirement. 2. Control records must guarantee product/process traceability and conformity and are archived according to customer traceability requirements. 3. Results are reviewed, out-of-control points are noted. 4. Actions are implemented in accordance with reaction plan defined in Control Plan. 5. Results are analyzed for continuous improvement activities.	• All the results recorded where Control Plan requires. Records are readable. • Out of control points are noted, necessary action implemented. • Track back records to check results are archived, traceability ensured.
The Information Systems are managed in order to secure all the data related to complete material flow of product.	1. A list of Critical Information systems is established and managed. 2. Resources are identified, qualified and available during all production period. The responsibilities are defined for each critical Information system: • to guarantee the availability and the integrity of the data, • to ensure the breakdown service in case of failure, • to implement needed functional evolutions. 3. There are procedures for backup and recovery of the data, and they are regularly tested. 4. For all main changes concerning critical information systems, a qualification process is carried out prior to full implementation.	• List of critical systems (system connected to the material flow management, software, programs, system related to the traceability of the parts, to incoming, to shipping). • Resources are available on-site to manage IT, and experts on short term. • Check back-up plans for loss of data / computer breakdown / network breakdown. • Specific procedure for degraded modes.
Product audits (including dock audit as well) are defined and performed according to Control Plan.	1. Product audit procedure is defined and performed according to Control Plan, and all the results are documented. 2. Product audits completely cover KPC verification.	• Control Plan defines product audit clearly. • Check Product last audit results. Was it performed according to the required frequency? • In case of any NOK results, action taken.

(Continued)

TABLE 9.15 (*Continued*)
An Overview of Criteria for Preparation of an Audit

Required	Criteria of Requirements	Verification Possibilities
	3. Product audit is considered as a standardized work. 4. Decision criteria to alert the customer are defined. 5. In case of outsourced test, labs must be certified.	
Safety stock location, organization, and management comply with customer expectations.	1. Safety stock warehouse (SSF) is separate from the supplier site. The distance between the two is sufficient to protect against social risks, flood, fire, *etc*. The location is confidential and unknown to the personnel of the site. 2. The SSF is managed by staff independent from the supplier, qualified to storage and shipping supplier standards. Schedules of activity of the SSF warehouse are consistent with customer sites' activity. 3. A procedure of emergency releasing is defined and applied. The time of provision of the parts (washing, repackaging) is managed. A releasing is possible during the night or the weekend. 4. The rebuilding of the SSF (in case of releasing) and the inventory turnover are organized and tracked. On customer request, the level of safety stock can be communicated and the turnover frequency can be adjusted. 5. Safety stock is fully integrated in the management system of supplier stock.	• Stock levels. • Results of audit and inventories. • Stock conditions. • List of contacts with the customer plant. • What happens if there is a request for an emergency release

[**Strong warning**: Audit and inventory are not the same. An audit is a methodology to verify that the system is working as intended. Inventory, on the other hand, is what is physically on the facility. That is, a list of what is available in your premises.]

10 Selected Useful Methodologies That Can Help Improve Customer Service and Voice of the Customer

OVERVIEW

So far, we have discussed critical and essential approaches to CS, VOC, and continual improvement with the goal of doing the right job right the first time and on time. In this chapter, we have addressed some additional methodologies that organizations may follow toward this end. They are: (a) production system, (b) RSEA. (c) warranty, (d) Weibull distribution, (e) FMEA, (f) Six sigma, (g) systems engineering, (h) APQP, (i) PPAP, (j) Capacity, and (k) OEE. We do not want to imply that these are not essential to CI, but rather we want to sensitize the reader that there are many other methodologies that a given organization may choose on the way to improvement and CS. The reason why we have included these topics separately is because in most organizations, they are part of the topics that we have chosen to discuss at length. On the other hand, they are important enough for any organization to be aware of their existence and apply them as necessary.

A TYPICAL TOTAL PRODUCTION SYSTEM (TPS)

In today's competitive world of quality, an integrated assessment tool must exist to evaluate and drive deeper into specific elements and processes, the notion of continual improvement in the entire organization (vertical and horizontal awareness by management and operator personnel).

One of the oldest and comprehensive TPS was developed by Ford Motor Company. However, one of the most talked about TPS is the one introduced by Toyota Corporation. Other companies have developed their own. Here, we present a generic summary of a typical TPS based on both Ford's and Toyota's systems. This outline is based on ten fundamental principles. They are:

- Total productive maintenance
- Training
- In-station process control

DOI: 10.1201/9781003650614-10

- Manufacturing engineering
- Environmental
- Industrial materials
- Safety and Health Achievement Recognition Program (SHARP): SHARP is a program that recognizes small employers who operate an exemplary safety and health management system. Acceptance into SHARP by OSHA is an achievement of status that will single you out among your business peers as a model for worksite safety and health and will reap rewards for your business. Upon receiving SHARP recognition, your worksite will be exempt from programmed inspections during the period that your SHARP certification is valid.
- Work groups
- Managing
- Synchronous material flow (SMF)

The Ford's Approach to TPS

The Ford Production System (FPS) was created to standardize and add stability to the existing manufacturing processes. In essence, this program is to focus on zero defects and 100%throughput. More specifically, by utilizing policy deployment, visual management, process confirmation, and time and data management, this process will deliver a more aligned and capable organization, one that will be continually improving and ultimately building the best manufacturing environments in the world. The metrics for this program are based on the **S**afety, **Q**uality, **D**elivery, **C**ost, **P**eople, **M**aintenance, and **E**nvironment (SQDCPME).

By measuring these metrics, Ford Motor Company has recognized that they are the pillars of their CI process, especially in their manufacturing operations. These pillars are aiming to continually improve their processes through standardization and *andon* recognition. In addition, these metrics seek to create stability to the inputs, normalize the processes, and seek change (if necessary) for the outputs. Specifically, the FPS is based on the following categories:

- *Continual improvement Board*: It is used to visually manage the review and follow-up of elevated issues of the work group.
- *Start-up confirmation*: It provides a time where the team leader makes sure that all operators are OK to start building. If an operator identifies a problem, it is immediately solved or escalated to the supervisor.
- *Results process/support process*: These are working meetings to analyze and diagnose workstation and process results. The intention is to make problems evident and make improvements.
- *Time and data management*: It is designed to assist the plant in identifying and eliminating waste through improved planning and alignment.
- *Basic administration*: It is another tool to help reduce waste by clearly defining individual roles and responsibilities, allowing time to change culture and focus on the teaching and the coaching of the standards necessary for improvement.

- *Kaizen*: As problems are escalated to the CI board and the solution is not readily evident, they can be further escalated into a Kaizen project. If the problems are complicated, then the Six Sigma methodology may be employed.
- *Standardized work*: The focus of standardized work is on standard process adherence and creating an ideal workstation, process, and method for each operation. This is a three-phase process to accomplish it: (a) Job observation, (b) deep knowledge operator review, and (c) identification and understanding of abnormal conditions. During the last two phases of this process, the team leader will replace the operator, who will have an in-depth discussion with the supervisor offline to review the job standards.
- *Star points*: They are policies, procedures for new processes that establish approaches to further promote work team involvement based on things learned from current practices.

In the final analysis, the Total Production System is: (a) People working together using TPS tools, (b) continual improvement in SQDCPME, and (c) focus on eliminating waste. It IS NOT: (a) Another modern operating agreement (MOA), (b) in conflict with the master labor agreement (if it exists), (c) in conflict with local labor agreements (if they exist), and (d) focused on eliminating people. In other words, the TPS is focused on a philosophy that encourages people's participation and *customer satisfaction*. Even though there are no rigid requirements, the philosophy is based on three fundamental output ideas. They are:

1. To provide the customer with the highest quality products, at the lowest possible cost, in a timely manner with the shortest possible lead times.
2. To provide employees with work satisfaction, job security, and fair treatment.
3. It gives the company flexibility to respond to the market, achieve profit through cost reduction activities, and long-term prosperity.

These fundamentals are indeed striving for the absolute elimination of waste, overburden, and unevenness in all areas to allow employees to work smoothly and efficiently. This is so because they are built (a) on standardization to ensure a safe method of operation and (b) to provide a consistent approach to quality. By doing so, the organization and employees *always* seek to continually improve their standard processes and procedures in order to ensure maximum quality, improve efficiency, and eliminate waste, by emphasizing the following simple concepts:

- *Kaizen*: It is the heart of the TPS. It is ingrained in the employee's minds that continual improvement in all that people and machines do, that can be improved continually in small steps.
- *Just-In-Time*: It consists of allowing the entire production process to be regulated by the natural laws of supply and demand. Customer demand stimulates the production of a product. In turn, the production of the product stimulates production and delivery of the necessary parts and so on. The result is that the right parts and materials are manufactured and provided in

the exact amount needed – and when and where they are needed. This process is regulated with the *Kanban* method. The Kanban method depends on a card which is attached to every component that is removed from a process and returned when the component is used. The return of the Kanban to its source stimulates the automatic re-ordering of the component in question. This Kanban process minimizes paperwork, maximizes efficiency, and allows the employees of the process to be completely in charge
- *Jidoka*: A Japanese word that means automation with a human touch. The idea here is to design machinery in such a way that if a problem arises, the machine will stop and call attention to the problem immediately. The *Jedoka* principle is a central concept in any application of the TPS.

Suggested Implementation of a Production System

A production system, by definition, is a lean, flexible, and disciplined common production system defined by a set of principles and processes that employs groups of capable and empowered people learning and working safely together in the production and delivery of products that consistently exceed customers' expectations in quality, cost, and time. What makes this system so powerful is that it focuses on *customer satisfaction* through some very important concepts, such as:

- Stability which indicates consistency, repeatability, and predictability.
- Continuous Flow which indicates production without interruptions.
- Synchronized production which indicates parallel production.
- Pull system which indicates a dynamic inventory system such as Kanban.
- Level Production which indicates a predictable production.

To implement these concepts, there are many options for every organization to have its culture and expectations. However, a good start is to begin by thorough plant visits and reviews to identify *what* is now. Typical things to consider are:

- *Visit application areas*: Talk with employees about what their role in the implementation of a production system. Explain the expectations.
- *Take a waste walk*: Observe areas where waste is prevalent by walking through the processes.
- Conduct frequent (at least monthly) meetings between senior leadership and plant manager to review the work plan, progress/timing, and provide coaching on how to deal with the implementation issues. A good start here is the QOS (Quality Operating System) meeting.
- *Reinforce the importance of all employees in getting involved and support*: Make sure that employees do not become suspicious of the organization's motives to eliminate them.
- *Develop a material flow plan*: For example: For current issues, produce the number of scheduled units each day at the lowest assembly variable cost with the highest quality. For future issues, build a customer-driven product sequence in a predictable and stable manner at the lowest total cost and time

with the highest quality. Understand and communicate to everyone that this new approach, new vision, perspective, goal, system requires less of everything – time, space, effort, capital – to produce goods and services with fewer mistakes in smaller production volumes. It is another way of looking at Lean manufacturing. In doing so, consider the following:

- Defining the direction and support to the plants on implementation issues.
- Defining measurement and reward for success.
- Defining appropriate and applicable measurables.
- Making sure that the appropriate frequency for meetings is established.
- Reassuring employees that their jobs are safe and not in jeopardy because of the implementation of the new system.
- Making the production system a priority in the plant(s) and not another flavor of the month initiative. Offer encouragement and support whenever needed.
- Making the total cost a key measurable in the plant's performance review.
- Defining appropriate milestones for reviews and assessments.
- Assigning a committee to lead the implementation process.

SUPPLIER REQUEST FOR ENGINEERING APPROVAL (SREA)

We all know that what is designed is not always what manufacturing can deliver. It is possible, and in fact, it happens quite often when the supplier finds a problem with the design. It is imperative at this point not to change the design on their own – even though they may have the best of intentions – but to notify the customer. The process of notification is called the SREA.

A SREA is an external supplier request to obtain customer approval for a plan to implement a proposed external supplier-initiated manufacturing process or site change. It is worth noting here that a SREA request for approval does not mean that the supplier can ship parts/products to the customer. For shipment, a PSW (part submission warrant) approval is required per PPAP requirements. The SREA is a way to prevent external supplier-initiated changes that adversely affect the product quality or manufacturing operations of the customer without their knowledge.

In its simplest form, the SREA is the official communication tool used by the customer to meet the external supplier-initiated change notification requirement of PPAP. Changes made without customer's review and approval have the potential to impact product quality, customer's plant processes, and more importantly, customer satisfaction. In essence, the SREA is a method of making sure that the customer does not receive any surprises from their supplier. If they do the unauthorized changes can result in unanticipated repair, replacement, quality cost, and late deliveries.

A SREA may be initiated everywhere the business is conducted and when an external supplier needs to make a change to a production manufacturing process, site/location, or part design after PPAP approval, or when an external supplier needs to make any change to a service-unique part/product or process after PPAP approval.

Generally, a SREA is required when manufacturing process and or site location changes to the original PPAP-approved process affecting:

- Site location
- Site construction affecting process or process flow
- Part/product processing
- Equipment changes
- Process flow
- Sub-supplier changes affecting site location, processing, process flow or materials
- Material changes within the existing design specification
- Material source changes
- Production

In more specific terms, an SREA is required for design changes affecting:

- Part/product specifications
- Part/product cost
- Appearance, quality, or performance of end-item characteristics
- Compliance with government regulations
- Function, fit or appearance of component/subsystem/system within a total vehicle system
- Interchangeability of end items
- Serviceability of the part/product
- Subsystem Design Specifications, Worldwide Standards and Specifications (if applicable)
- Product Assembly Plant operations
- Warranty and reliability

Service unique products are parts/products released by the customer and not common with production parts/products (*i.e.,* remanufactured components, service chemicals, *etc.*), and past model parts/products no longer used in production or for original equipment warranty repairs, but still produced for service. The SREA requirements for service unique manufacturing and site location changes are the same as the list above under the general requirements. Sometimes different customers will have a different department handling current model service unique parts/product from the older models. Service-unique design changes that require an SREA are: Any supplier-initiated changes outlined in the above list of general requirements, including:

- Change components of an assembly due to availability issues (obsolete, not currently available, tooling down, *etc.*).
- Ship without a component of an assembly due to availability issues.
- Change the content of a kit due to availability issues.
- Substitute essentially identical final assemblies.
- Expand to coverage of one part/product and drop another for complexity reduction.

- Material changes or substitutions resulting from obsolete or prohibited material revisions.
- Worn tooling dimensional and tolerance change approval.
- Request to produce part/product to meet last run samples – as a result of:
    - Loss of Check Fixtures
    - Loss of Tools
    - Loss of Master Sample parts/product
- Design Record not up to date due to changes made with no record, changes made with an incomplete record, changes made with an inaccurate record
- Drawing not up to date for all of the above
- No Drawing
- No formal Design Record

Temporary SREA

Sometimes the supplier will request to ship parts/product made to a temporary engineering specification. This request is the same as the regular SREA. The only difference is that the regular one is permanent, and this one is temporary. Reasons for consideration for a temporary SREA may be:

- Part/products temporarily made using an unapproved manufacturing process or site/location (at Tier 1 or sub-supplier).
- Temporary emergency change to the part/product or manufacturing process or site/location.
- Part/products temporarily not meeting customer's specifications.

Supplier Risks for Not Following a SREA

By not following the correct approach for an SREA, the risk to the supplier may be at least in the following areas:

- Loss of customer quality system, *i.e.,* loss of Quality status through violation of trust.
- Increase field warranty or customer plant containment costs (through third party inspection or some other method of segregating good from bad parts).
- Lost production and reinstatement of the prior process.
- Loss of "good will."
- Increase amount of customer dissatisfaction. (This is a common source of recalls.)
- Loss of business (current and future).

WARRANTY

In the world that we live in, we have become a very litigious society. We look for perfection, and any deviation from that perfection can cause trouble for the OEM customer but also for the supplier. To mitigate the failure of a product, organizations all over the world have initiated warranties that cover their specific products or services. These warranties may involve the ultimate customer (consumer) or the supplier and their sub-suppliers. Therefore, for any warranty program to work effectively, there are two approaches that must be considered. They are:

1. Issues that involve common cause processes. Here, the focus is on (a) warranty reduction programs and (b) warranty chargebacks.
2. Issues that involve special cause processes. Here, the focus is on (a) warranty spikes, (b) field service action cost recovery, and (c) warranty cost reduction process once the product is in production.

The idea of a warranty program for the customer is to establish direct responsibility for failures to any external supplier for all or a portion of any costs associated and validated with warranty claims. For the supplier, it is important to negotiate an appropriate and applicable contract to pay claims under the rubric of a warranty claim. This negotiation is best conducted as early as possible, preferably at sourcing.

WEIBULL DISTRIBUTION

The Weibull distribution is a good distribution to understand and evaluate life data analysis. Weibull and Log Normal analysis are particularly good for failure analysis. Because of this ability to evaluate failures, there are new applications of this distribution in medical and dental implants, warranty analysis, life cycle cost, materials properties, and production process control. Of course, the use in the automotive, defense, aerospace, and other mature industries continues to be high. Other methods similar to Weibull exist, but they are for specific applications. Some related quantitative models that the reader may be interested in, such as the binomial, Poisson, Kaplan–Meier, Gumbel extreme value, and the Crow-AMSAA, are worth considering. A very good, detailed introduction to Weibull is Abernethy (2009) and a summary of it with several examples is Stamatis (2010 chapter 15).

The primary advantage of Weibull analysis is the ability to provide reasonably accurate failure analysis and failure forecasts with extremely small samples. Solutions are possible at the earliest indications of a problem without having to "crash a few more." Small samples also allow cost-effective component testing. For example, "sudden death" Weibull tests are completed when the first failure occurs in each group of components (say, groups of four bearings). If all the bearings are tested to failure, the cost and time required are much greater. Another advantage of Weibull analysis is that it provides a simple and useful graphical plot of the failure data. There are several software packages in the market to calculate appropriate reliability values as well as plotting the distributions. One of those packages is ReliaSoft's Weibull++ software, which provides a complete array of life data analysis tools and is very friendly to use. Minitab is also good for defining and plotting the distribution functions via linear regression, which can also be used to numerically assess goodness of fit and estimate the parameters of the Weibull distribution. The gradient informs one directly about the shape parameter, κ and the scale parameter ʎ can also be inferred. The simplest of all methods to identify a distribution is to construct a histogram, for it will show the shape of the distribution. Using the Minitab software to identify the distribution, go to Stat > Quality Tools > Individual Distribution Identification in Minitab. This handy tool allows you to easily compare how well your data fit 16 different distributions. It produces a lot of output both in the session window and graphs, but don't be intimidated. To appreciate the output of the Minitab, there are three measures you need to know.

1. *Anderson-Darling statistic (AD)*: Lower AD values indicate a better fit. It's generally valid to compare AD values between distributions and go with the lowest.
2. *P-value*: You want a high p-value. A low p-value (*e.g.*, < 0.05) indicates that the data does not follow that distribution. For some 3-parameter distributions, the p-value is impossible to calculate and is represented by asterisks.
3. *LRT P*: For 3-parameter distributions only, a low value indicates that adding the third parameter is a significant improvement over the 2-parameter version. A higher value suggests that you may want to stick with the 2-parameter version.

The term life data refers to measurements of product life. Product life can be measured in hours, miles, cycles, or any other metric that applies to the period of successful operation of a particular product. Since time is a common measure of life, life data points are often called "times-to-failure," and product life is described in terms of time throughout the life of the part/product. There are different types of life data, and because each type provides different information about the life of the product, the analysis method will vary depending on the data type. For example: (a) With "complete data," the exact time-to-failure for the unit is known (*e.g.*, the unit failed at 100 hours of operation). (b) With "suspended" or "right censored" data, the unit operated successfully for a known period of time and then continued (or could have continued) to operate for an additional unknown period of time (*e.g.*, the unit was still operating at 100 hours of operation) (c) With "interval" and "left censored" data, the exact time-to-failure is unknown but it falls within a known time range (*e.g.*, the unit failed between 100 hours and 150 hours (interval censored) or between 0 hours and 100 hours (left censored)). Life data analysis requires the following steps to be effective:

- Gather life data for the product.
- Select a lifetime distribution that will fit the data and model the life of the product.
- Estimate the parameters that will fit the distribution to the data. Once this information is collected, then the analysis begins. Typical outcomes are to: (a) Generate plots and results that estimate the life characteristics of the product, such as the reliability or mean life. (b) Compare suppliers or designs based on reliability.
- Demonstrate that an item meets specified reliability.

- Make predictions about performance during the useful life (or warranty) period.
- Use plots and other reports to effectively communicate performance estimates to management.

Once you have calculated the parameters to fit a life distribution to a particular data set, you can obtain a variety of plots and calculated results from the analysis, including:

- *Reliability given time*: The probability that a unit will operate successfully at a particular point in time. For example, there is an 80% chance that the product will operate successfully after 3 years of operation.
- *Probability of failure given time*: The probability that a unit will fail at a particular point in time. Probability of failure is also known as "unreliability," and it is the reciprocal of the reliability. For example, there is a 10% chance that the unit will fail after 3 years of operation (probability of failure or unreliability) and an 80% chance that it will operate successfully (reliability).
- *Mean Life*: The average time that the units in the population are expected to operate before failure. This metric is often referred to as "mean time to failure" (MTTF) or "mean time before failure" (MTBF).
- *Failure rate*: The number of failures per unit time that can be expected to occur for the product.
- *Warranty time*: The estimated time when the reliability will be equal to a specified goal. For example, the estimated time of operation is 3 years for a reliability of 90%.
- *BX Life*: The estimated time when the probability of failure will reach a specified point (X%). For example, if 10% of the products are expected to fail by 3 years of operation, then the B10 life is 3 years. (Note that this is equivalent to a warranty time of 3 years for a 90% reliability.) The B10 is usually the common statistic used; however, the "X" can be any number.
- *Probability plot*: A plot of the probability of failure over time. (Note that probability plots are based on the linearization of a specific distribution. Consequently, the form of a probability plot for one distribution will be different than the form for another. For example, an exponential distribution probability plot has different axes than those of a normal distribution probability plot.)
- *Reliability vs. Time Plot*: A plot of the reliability over time.
- *pdf Plot*: A plot of the probability density function (pdf). It plots a function, usually probability density (pdf) or cumulative distribution function (cdf), over an interval containing the "interesting" part of the function. The interval is based on quantiles computed from a supplied cdf or quantile function.
- *Failure rate vs. time plot*: A plot of the failure rate over time.
- *Contour plot*: A graphical representation of the possible solutions to the likelihood ratio equation. This is used to make comparisons between two different data sets.

Is Weibull always the best choice? The answer is: *definitely not*. However, it is appropriate for survival analysis, reliability engineering, failure analysis, and industrial engineering to represent manufacturing and delivery times, and in evaluating extreme value theory. On the other hand, it is not appropriate for product failures caused by chemical reactions or corrosion, which are usually modeled with the lognormal distribution. The following are typical examples of engineering problems solved with Weibull analysis:

- A project engineer reports three failures of a component in service operations during a three-month period. The Program Manager asks, "How many failures will we have in the next quarter, six months, and year?" What will it cost? What is the best corrective action to reduce the risk and losses?
- To order spare parts and schedule maintenance labor, how many units will be returned to the depot for overhaul for each failure mode, month-by-month next year? The program manager wants to be 95% confident that he will have enough spare parts and labor available to support the overall program.
- A state Air Resources Board requires a fleet recall when any part in the emissions system exceeds a 4% failure rate during the warranty period. Based on the warranty data, which parts will exceed the 4% rate and on what date?
- After an engineering change, how many units must be tested for how long, without any failures, to verify that the old failure mode is eliminated, or significantly improved with 90% confidence?
- An electric utility is plagued with outages from superheater tube failures. Based on inspection data, forecast the life of the boiler based on plugging failed tubes. The boiler is replaced when 10% of the tubes have been plugged due to failure.
- The cost of an unplanned failure for a component, subject to a wear-out failure mode, is twenty times the cost of a planned replacement. What is the optimal replacement interval?

FMEA

Failure Modes and Effects Analysis (FMEA) is a bottom-up method of analyzing a design or a process, and is widely used in many industries. In the USA, automotive companies such as Chrysler, Ford, and General Motors require that this type of analysis be carried out. There are many different company and industry standards, but one of the most widely used is the Automotive Industry Action Group (AIAG) model. Using this standard, you start by considering each component or functional block in the system and how it can fail, referred to as failure modes. You then determine the effect of each failure mode and the severity on the function of the system. Then you determine the likelihood of occurrence and of detecting the failure. The procedure is to calculate the Risk Priority Number, or RPN, using the formula: RPN = Severity × Occurrence × Detection.

Although this RPN has been used and continues to be used by many organizations, a better way to address risk priority is based on the numerical value of Severity (priority one); followed by Severity × Occurrence (priority two); followed by the value of the RPN. Some organizations define priority as Severity (9–10); followed by severity (5–8); followed by the RPN. **In all cases, there should never be a threshold of an RPN value**.

The second stage is to consider corrective actions that can reduce the severity or occurrence, or increase detection. Traditionally, you start with the higher RPN values, which indicate the most severe problems, and work downwards. The RPN is then recalculated after the corrective actions have been determined. The intention is to get the RPN to the lowest value. If, on the other hand, the organization has chosen to evaluate the FMEA based on Severity, Severity x Occurrence, RPN, the focus will be to make sure that severity is controlled or changed with an appropriate design change. This will be followed by redundant systems to minimize the occurrence, followed by appropriate controls to catch the failure before it reaches the customer. For detailed information on FMEAs, see Stamatis (2003, 2015).

SIX SIGMA

Developed by Motorola in the 1980s, this improvement methodology was created to reduce errors, waste, and variations, and increase quality and efficiency in manufacturing. It has since been adapted for use in other types of business processes, and is today in practice in some of the top companies around the globe, including Motorola, General Electric, Siemens, Honeywell, Ford, Allied Steel, and many more. Six sigma-driven companies use data to examine, manage, and enhance operational performance by eliminating and preventing flaws in goods and related processes, such as design, management, production, consumer satisfaction, and service delivery (Pande et al, 2000; Stamatis 2002–2003; Harry and Schroeder 2006).

The bulk of the work in a six-sigma project would be in defining failures, measuring deviations, and other activities that ultimately lead to product quality. In fact, Six Sigma is used as a term for a management style, with the ultimate goal of high levels of customer satisfaction (Stamatis, 2004). Eckes (2001) and Pyzdek (2001) have suggested that perhaps the most objective way of looking at Six Sigma is to recognize that the Six Sigma methodology essentially provides a framework, and most importantly, a strongly branded corporate initiative, for an organization to:

- Train its people to focus on key performance areas.
- Understand where the organization wants to go (its strategy, related to its market-place).
- Understand the services that the organization's customers need most.
- Understand and better organize the main business processes that deliver these customer requirements.
- Measure (in considerable detail) and improve the effectiveness of these processes.

TABLE 10.1
Relationship of Sigma and Defect

Sigma Level	Defects per Million	Defects Percentage
1	691,462	69%
2	308,536	31%
3	66,807	6.7%
4	6,210	0.062%
5	233	0.023%
6	3.4	0.00034%
7	0.019	0.0000019%

In specific terms, the Six Sigma methodology at its core centers on the following concepts:

- *Critical to quality*: Attributes most important to the customer.
- *Defect*: Failing to deliver what the customer wants.
- *Process capability*: What your process can deliver.
- *Variation*: What the customer sees and feels.
- *Stable operations*: Ensuring consistent, predictable processes to improve what the customer sees and feels.
- *Design for Six Sigma*: Designing to meet customer needs and process capability.

One of the guiding principles behind the methodology is that variation in a process creates waste and errors. Eliminating variation, then, will make that process more efficient, cost-effective, and error-free. This may sound like a relatively straightforward concept, but its application in a complex and highly integrated business environment can be far from simple. The term sigma refers to a scale of measurement of quality in processes such as manufacturing. When using this particular scale, Six Sigma equates to just under 3.4 defects per million opportunities (DPMO). Table 10.1 shows the relationship between sigma and defects.

Six Sigma Methodology

Whether you are trying to implement Six Sigma throughout the entire organization or using it to improve the performance of your own department, you need to understand the two main Six Sigma categories. The first of these is DMAIC, which stands for: define, measure, analyze, improve, and control. The second is DMADV, which represents: define, measure, analyze, design, and verify.

DMAIC is used to apply the principles of Six Sigma to existing business processes. For instance, if you are trying to find out how to make a particular process more effective, you would use DMAIC to break down the process into its component

TABLE 10.2
Typical Tools Used in the DMAIC Model

Define Tools	Measure Tools	Analyze Tools	Improve Tools	Control Tools
QFD Surveys GQRS Warranty In-plant data Pareto	Blueprints Process Sheets Metrics Flow Charts Linkage Matrices FMEAs Measurement System Analysis Capability Analysis	Multivariate Analysis SPC & Charting Basic Stats - Hypothesis - Basic DOE Check Sheets Histograms Scatter Diagrams Regression	Designed Experiments - Full Factorial - Fractionals - Comparison Measurement System Analysis Capability Statistical Tolerancing of Process	Control Plan Operator Illustrations Operator Instructions Error Proofing Automated Control QPS Variation Monitoring

parts. Using this Six Sigma model, you would start by defining the problems and project goals, measuring data relating to the current process, and analyzing your findings to identify cause-and-effect relationships. The next step involves improving existing processes based on your data analysis. Finally, you need to implement controls to avoid variation in the process going forward. Typical tools used on each step are shown in Table 10.2.

On the other hand, the classical design for Six Sigma (DFSS) follows the DMADV model, which shares some steps in common with the DMAIC model. It is used for Six Sigma projects that create new product or process designs. Using the DMADV model, you would start by defining your design goals. The next step is to measure the required quality characteristics, product or production process capabilities, and associated risks. You would then conduct an analysis of your findings to develop an appropriate solution. After that, you are ready for the design phase of the new product or process. Once the design is complete, you must test it and verify that it works. Following these steps will result in a successful Six Sigma implementation. Each of the acronyms is defined very precisely in the following manner:

- *Define*: Address customer needs in relation to a product or service.
- *Measure*: Involve the use of electronic or mechanical, or physical data collection to measure customer needs, response to product, or review of services.
- *Analyze*: Utilize metrics to evaluate areas where the product or service can be better aligned to customer goals and needs.
- *Design*: Overlap the improvement of business processes that streamline corporate goals to best meet client and customer needs.

- *Verify*: Build a system of tests and models to check that customer specifications are being met through ongoing improvements.

In the evolution of the classical DFSS process, there is also the DCOV model, which has gained in popularity and stands for:

- *Define*: Address customer needs in relation to the product or service desired.
- *Characterize*: Mathematically transform the customer's wants and needs into specifications or requirements.
- *Optimize*: Out of the possible outcomes, select the best alternative that will satisfy the customer. This is an issue of measuring strengths and weaknesses or risk and benefit.
- *Verify*: Verify that what you have selected is what the customer wants and is willing to pay for.

For a very detailed discussion and implementation, see Stamatis (2002–2003. 6σ – Vol. 1–VII).

SYSTEMS ENGINEERING

Systems engineering is an interdisciplinary approach to design, implementation, and evaluation that holds the key to the successful development of complex human-made systems. It can be used both by OEMs and suppliers to have effective designs.

The classic model follows the letter V (although there are variations of it) and is shown in Figure 10.1. The V-model provides guidance for the planning and

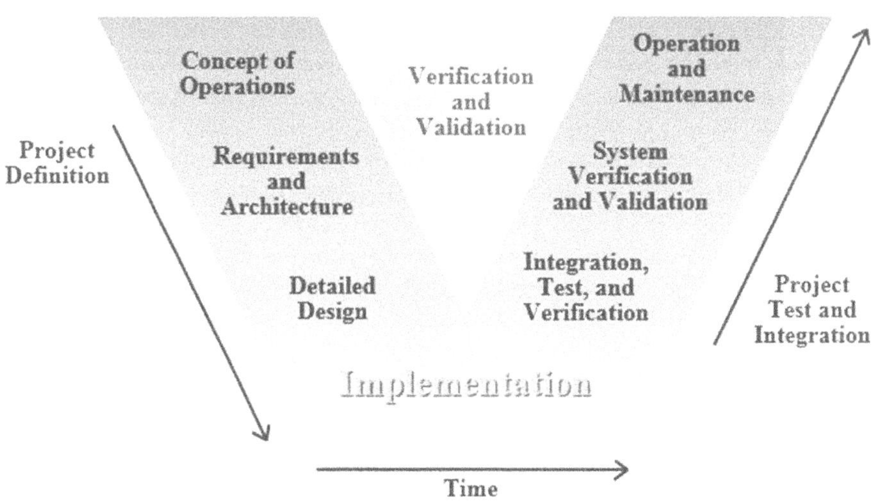

FIGURE 10.1 The typical systems engineering V-model.

realization of projects. The following objectives are intended to be achieved by the project execution:

1. *Minimization of project risks*: The V-model improves project transparency and project control by specifying standardized approaches and describing the corresponding results and responsible roles. It permits an early recognition of planning deviations and risks and improves process management, thus reducing the project risk.
2. *Improvement and guarantee of quality*: As a standardized process model, the V-model ensures that the results to be provided are complete and have the desired quality. Defined interim results can be checked at an early stage. Uniform product contents will improve readability, understandability, and verifiability.
3. *Reduction of total cost over the entire project and system life cycle*: The effort for the development, production, operation, and maintenance of a system can be calculated, estimated, and controlled in a transparent manner by applying a standardized process model. The results obtained are uniform and easily retraced. This reduces the requirements of dependency on the supplier and the effort for subsequent activities and projects.
4. *Improvement of communication between all stakeholders*: The standardized and uniform description of all relevant elements and terms is the basis for the mutual understanding between all stakeholders. Thus, the frictional loss between user, supplier, and developer is reduced.

Customer requirements ←——→ Customer satisfaction

It is very important to note the left leg of the V, which is design-oriented, and the right is manufacturing-oriented. So, generally, the model is depicted as a wide left leg with the arrow going back and forth to indicate adjustments and changes in the design. The right leg is much narrow to indicate the embedded changes in the design and, as a consequence, much easier manufacturing. The left side of the V-model begins with customer requirements, and the right finishes with customer satisfaction. What is very important about these two characterizations is that there is a double arrow that connects the two to signify continual sharing of information for both compliance as to what the customer is asking (requirements) and whether or not the customer is satisfied with what is received and paid for. Finally, it is also interesting to note that validation and verification are in the middle and generally are shown with a double arrow going back and forth between design and manufacturing. IEEE (2011, p. 452) defines validation and verification in very specific terms as: (a) *Validation*. The assurance that a product, service, or system meets the needs of the customer and other identified stakeholders. It often involves acceptance and suitability with external customers, and (b) *Verification*. The evaluation of whether or not a product, service, or system complies with a regulation, requirement, specification, or imposed condition. It is often an internal process. However, in simple terms, the essence of validation can be expressed by the question "are you building the right thing?" and verification by "are you building it right?"

To improve the design and manufacturing systems, engineering encourages the use of tools and methods to better comprehend and manage complexity in systems. Some examples of these tools are:

- System model, modeling, and simulation
- System architecture
- Optimization
- System dynamics
- Systems analysis
- Statistical analysis
- Reliability analysis, and
- Decision making

On the other hand, early in the development stage of the design, when the primary purpose of a systems engineer is to comprehend a complex problem, graphic representations of a system are used to communicate a system's functional and data requirements. Common graphical representations include:

- Functional flow block diagram (FFBD)
- Model-based design, for example, Simulink, VisSim, *etc.*
- Data flow diagram (DFD)
- N2 Chart
- IDEF0 diagram
- Use case diagram
- Sequence diagram
- USL function maps and type maps.
- Enterprise architecture frameworks, like TOGAF, MODAF, Zachman Frameworks, *etc.*
- Value engineering

Value engineering (VE) is a systematic method to improve the "value" of goods or products and services by using an examination of function. Value, as defined, is the ratio of function to cost. Value can therefore be increased by either improving the function or reducing the cost. It is a primary tenet of VE that basic functions be preserved and not be reduced as a consequence of pursuing value improvements.

The reasoning behind VE is as follows: If marketers expect a product to become practically or stylistically obsolete within a specific length of time, they can design it to only last for that specific lifetime. The products could be built with higher-grade components, but with VE they are not because this would impose an unnecessary cost on the manufacturer, and to a limited extent also an increased cost on the purchaser. Value engineering will reduce these costs. A company will typically use the least expensive components that satisfy the product's lifetime projections.

Value engineering is often done by systematically following a multi-stage job plan. Larry Miles' original system was a six-step procedure which he called the "value analysis job plan." Others have varied the job plan to fit their constraints.

Depending on the application, there may be four, five, six, or more stages. One modern version has the following eight functions:

1. Preparation
2. Information
3. Analysis
4. Creation
5. Evaluation
6. Development
7. Presentation
8. Follow-up

These eight functions may be formally presented in four basic steps in the job plan as:

1. *Information gathering*: This asks what the requirements are for the object. Function analysis, an important technique in value engineering, is usually done in this initial stage. It tries to determine what functions or performance characteristics are important. It asks questions like: What does the object do? What must it do? What should it do? What could it do? What must it not do?
2. *Alternative generation (creation)*: In this stage, value engineers ask: What are the various alternative ways of meeting requirements? What else will perform the desired function?
3. *Evaluation*: In this stage, all the alternatives are assessed by evaluating how well they meet the required functions and how great the cost savings will be.
4. *Presentation*: In the final stage, the best alternative will be chosen and presented to the client for final decision.

These four steps of VE follow a structured thought process to evaluate options as follows.

Gather information

- What is being done now?
- Who is doing it?
- What could it do?
- What must it not do?

Measure

- How will the alternatives be measured?
- What are the alternate ways of meeting requirements?
- What else can perform the desired function?

Analyze

- What must be done?
- What does it cost?

Generate

- What else will do the job? (Identify alternatives)

Evaluate

- Which ideas are the best? (Must have appropriate and applicable criteria.)
- Develop and expand ideas.
- What are the impacts? What is the cost? What is the performance? (Are the proposals meeting and or exceeding customer needs? Wants? Expectations? If so, to what degree?)
- Present ideas
- Sell alternatives (focus on the most successful alternative)

APQP

Overview

As we mentioned in Chapter 1, the principles of APQP are very important from a customer's perspective, for they assure timing, capacity, and quality characteristics being within the necessary requirements. In this chapter, we will focus on the requirements of APQP from the supplier's perspective.

Traditionally, APQP has been misunderstood and taken seriously very late in the process of launching. Suppliers do not take advantage of the positive benefits that a system such as the APQP can deliver and the many value-added goals that it presents when it is viewed as a process rather than a single tool to be checked in a list of multiple activities just because "the customer says so." APQP is a structured method of defining and establishing the steps necessary to assure that a product satisfies the customer. The goal then is to facilitate communication with everyone involved to make sure that all required steps are completed on time. It is imperative that any form of quality planning in any organization must *NOT* be viewed as a system to be used by certain individuals and for specific situations, but rather as a *process* that encompasses all aspects of a program, *i.e.*, from conception to completion. This commitment to planning for quality applies to everyone involved – from management to operators.

Effective advanced quality planning depends on a company's top management commitment to the effort required in achieving customer satisfaction. In essence, this effectiveness is a culture change and a mindset that calls for continual improvement and waste reduction.

For a typical supplier, the benefits of APQP are to:

- Promote early identification of required changes.
- Avoid late changes.
- Provide a quality product on time at an acceptable cost to satisfy the customer.
- Provide uniformity and standardization in the process of communicating and reporting any issues or problems to the customer.

APQP Initiatives

The APQP initiative is important in that it demonstrates the continuing efforts of both customer and supplier to become quality leaders in their respective industries. It does this by providing value-added goals explicitly focused on all pertinent activities to avoid concerns rather than focusing on the results in the product or service throughout all the phases, until delivery of the product or service to the customer.

This is the reason that the deliverables discussed earlier must be clear and understood by all concerned. In fact, to make the point, we suggest that the reader review Appendix A. The appendix is a typical detailed *expectation list* of a site assessment by the customer to the supplier's facility. By understanding the value-added goals, the supplier will benefit immensely because they add to better quality, lower cost, productivity increases, and timely delivery. Some of the value-added goals may become instrumental to:

- Reinforce the organization's focus on CI in quality, cost, and delivery.
- Provide the ability to look at an entirely new program as a single unit.
- Be prepared for every step in the creation of the product.
- Identify where the most amount of effort must be centered.
- Create a new product with efficiency and quality.
- Provide a better method for balancing the targets for quality, cost, and timing.
- Deploy targets that use detailed practical deliverables within specific timing schedule requirements.
- Provide a tool (methodology) for all concerned to follow up on all planning processes.
- Provide consistency and stabilization, resulting in cost improvement opportunities, including the reduction of special sample test trials.

Understanding the program requirements makes it easier for the supplier to fulfill the expectations of the customer, but it also provides the following advantages to the customer.

- Clarifies the expectations early on, in the relationship.
- Focuses on the deadlines to make sure they are met.
- Tracks special characteristics (critical and significant).
- Evaluates and avoids quality, cost, and timing risks.
- Clarifies any bottlenecks and concerns for all appropriate product specifications using a common control plan.

Fundamental Elements of APQP

In the automotive industry, there are several (generic) elements in the APQP process, and they are published in AIAG (2008b). Here, we focus on key items that the supplier must be aware of and complete. For an extensive discussion, see Stamatis (2019).

We begin with the APQP focus, which is twofold:

- Up-front quality planning. The earlier the planning, the less the risk of failure.
- Identifying key processes so that customers are satisfied by evaluating and accepting the output. Determining appropriate and applicable support for continual improvement.

To deliver on these two focus items, the APQP process is divided into five phases. They are:

- Plan and define program
- Product design and development verification
- Process design and development verification
- Product and process validation and production feedback
- Launch, assessment and corrective action

Within these phases, there are five major activities:

- Planning
- Product design and development
- Process design and development
- Product and process validation
- Production

Out of all the key items in the APQP, seven major elements stand out. They are:

- Understanding the needs of the customer.
- Having a proactive feedback and corrective action.
- Designing within the process capabilities.
- Analyzing and mitigating failure modes.
- Having appropriate and applicable verification and validation.
- Having design reviews.
- Having a process to control special (critical and significant) characteristics.

The reader by now may feel somewhat confused because of all the divisions we just covered. The reason for all these divisions is to make sure that the supplier understands why the APQP is so very important for them to make sure they complete on time and without any glitches. To further demonstrate the importance of each element, we will provide specific definitions that split each element into six separate areas. These areas are:

- *Definition*: Identifies the requirements for the element.
- *Expectations*: Defines the requirements for the element. This usually is given to the supplier by the customer.
- *Lead responsibility*: Defines and identifies the activities responsible for lead reporting. Activities that all others will support in the completion of the element.

- *Support function*: Defines and identifies the support activities that will provide input to and assist the lead responsibility.
- *Timing*: Identifies the initial and final milestone timing for the element. It is worth mentioning here that although the customer initiates the boundary of the timing milestones, it is the supplier that confirms the material delivery date (MDD). So, it is imperative that the supplier understands and agrees to the timetable. Furthermore, it is the responsibility of the supplier to completely understand the timing plan for each element and should be able to identify the period of each phase. If the supplier disagrees with the MDD, let the customer be aware as soon as possible.
- *Deliverables*: Indicates the items that must be completed during three separate time frames for the element. The three time frames are:
    - *Initial*: Deliverables that must be completed upon initiation of the element or before the element enters the intermediate phase.
    - *Intermediate*: Deliverables that must be completed during the process of the element. These deliverables may be started during the initial phase, but must be completed before the final phase.
    - *Final*: Deliverables that must be completed before the process moves on to the next element.

PPAP

Overview

The essence of a PPAP is to make sure that the supplier knows and monitors their process to the point where the product going to the customer is *flawless*. The PPAP is product-specific and *never* facility-specific. Specifically, any PPAP focuses on six primary areas. They are:

1. If a process owner exists.
2. If a process is defined.
3. If a process is documented.
4. If linkages of process are established.
5. If process is monitored, analyzed, and improved.
6. If records are maintained.

These six areas are embellished into 18 specific requirements. These are not new but repositioned requirements of the APQP process (AIAG, 2008b). This means that if the supplier has done diligence work on time for all the APQP requirements on their due time (milestone), the PPAP process is not difficult or time consuming. The 18 elements of the full PPAP are:

1. *Design records:* A printed copy of the drawing needs to be provided. If the customer is responsible for designing, this is a copy of the customer drawing that is sent together with the Purchase Order (PO). If supplier is responsible for designing, this is a released drawing in supplier's release

system. Each and every feature must be "ballooned" or "road mapped" to correspond with the inspection results (including print notes, standard tolerance notes, and specifications, and anything else relevant to the design of the part). Prints – if applicable, must be in the geometric and dimensioning tolerance (G & DT) format.

2. *Engineering change documents:* A document that shows the detailed description of the change. Usually, this document is called "Engineering Change Notice," but it may be covered by the customer PO or any other engineering authorization. If a change is initiated by the customer during the PPAP process, it is called an alert and is good for 60 days. If during these 60 days nothing happens, it reverts back to the original status. If the change is complete, then the appropriate documentation is updated and made part of the final PPAP. If it is a work in process, it may be closed and a new one issued, or the current one may be extended. (Obviously, the days may depend on the customer.)
3. *Engineering approval:* This approval is usually the engineering trial with production parts performed at the customer plant. A "temporary deviation" is usually required to send parts to the customer before PPAP. Customer may require other "Engineering Approvals."
4. *Design FMEA:* A copy of the Design Failure Mode and Effect Analysis (DFMEA) is reviewed and signed off by supplier and customer. If the customer is the design responsible, usually customer may not share this document with the supplier. However, the list of all special (critical and significant) or high-impact product characteristics should be shared with the supplier, so they can be addressed on the PFMEA and Control Plan. If the supplier is design responsible, then the FMEA must be shared with the customer. If the part is confidential, the customer may review it in the supplier's facility, but the customer cannot remove the document from the facility. It cannot be faxed, emailed, copied, or in any other way transferred unless properly authorized by the supplier.
5. *Process flow diagrams:* A copy of the Process Flow, indicating all steps and sequence in the manufacturing process, including incoming components, must be available for review.
6. *Process FMEA:* A copy of the Process Failure Mode and Effect Analysis (PFMEA) is reviewed and signed off by supplier and customer. The PFMEA follows the Process Flow steps, and indicates "what could go wrong" during the manufacturing and assembly of each component. All special characteristics must be accounted for and linked to the control plan.
7. *Control Plan:* A copy of the Control Plan is reviewed and signed off by supplier and customer. The Control Plan follows the PFMEA steps, and provides more details on how the "potential issues" are checked in the incoming quality, assembly process, or during inspections of finished products.
8. *Measurement system analysis studies:* MSA usually contains the Gage R & R (repeatability and reproducibility) for the special or high-impact characteristics, and a confirmation that gauges used to measure these characteristics are appropriate and calibrated.

9. *Dimensional results:* A list of every dimension noted on the ballooned drawing. This list shows the product characteristic, specification, the measurement result(s) and the assessment showing if this dimension is "OK" or "not OK." Usually, a minimum of six pieces is reported per product/process combination.
10. *Records of material/performance test results:* A summary of every test performed on the part. This summary is usually in the form of DVP&R (Design Verification Plan and Report), which lists each individual test, when it was performed, the specification, results, and the assessment pass/fail. If there is an engineering specification, usually it is noted on the print. The DVP&R shall be reviewed and signed off by both customer and supplier engineering groups. The quality engineer will look for a customer signature on this document. In addition, this section lists all material certifications (steel, plastics, plating, *etc.*), as specified on the print. The material certification shall show compliance with the specific call on the print.
11. *Initial process studies:* Usually, this section shows all Statistical Process Control (SPC) charts affecting the most critical characteristics. The intent is to demonstrate that critical processes have stable variability and are running near the intended nominal value. Sometimes, in this element, an *Initial Sample Inspection Report* will be attached to identify the material samples that were initially inspected before prototype stage.
12. *Qualified laboratory documentation:* Copy of all laboratory certifications (*e.g.,* A2LA, TS) of the laboratories that performed the tests reported on element 10. All laboratory tests must have traceability to NSIT or be approved by the customer.
13. *Appearance approval report:* A copy of the AAI (Appearance Approval Inspection) form signed by the customer. Applicable for components affecting appearance only.
14. *Sample production parts:* A sample from the same lot of the initial production run. The PPAP package usually shows a picture of the sample and where it is kept (customer or supplier).
15. *Master sample:* A sample signed off by customer and supplier, that is usually used to train operators on subjective inspections, such as visual or noise attributes.
16. *Checking aids:* When there are special tools for checking parts, this section shows a picture of the tool and calibration records, including dimensional report of the tool.
17. *Customer-specific requirements:* Each customer may have specific requirements to be included in the PPAP package. It is a good practice to ask the customer for PPAP expectations before even quoting for a job. For example, North America auto makers OEM (Original Equipment Manufacturer) requirements are listed on the IATF website.
18. *Production part approval process submission warrant (PSW):* This is the form that summarizes the whole PPAP package. This form shows the reason for submission (design change, annual revalidation, *etc.*) and the level of documents submitted to the customer. There is a section that asks for "results meeting all drawing and specification requirements: yes/no" refers

to the whole package. If there is any deviation, the supplier should note on the warrant or inform that PPAP cannot be submitted. *A successful launch is where the ramp up curve is met without missing parts or shortages and without the requirement to retrofit any component for quality, capacity or engineering reasons.* Here, we must make a very strong point, and that is. Even though a PSW may be completed and approved, the customer must approve the shipping for production purposes. That approval comes after "functional testing" at the customer's production facility. If the functional testing passes, then production is commenced; if not, then the parts of the PPAP that fail must be redone over again.

The notion of the PSW is very critical, and many suppliers do not understand the ramifications. Therefore, let us look at this PSW a little more in detail. We start with the scope of the PSW, which is to redefine both the interpretation and execution of the PSW and PPAP. Due to "compromised" interpretations because of timing pressures, launches and Job1 have often been disrupted and delayed due to parts that either do not meet specification or do not meet the required volume. This is a very serious problem, and every OEM is very concerned about it.

Therefore, to understand the importance of the PSW, we must understand and internalize "what is," and "what is not" a PSW. A PSW run is conducted at the intended point of production using, the production machinery, tools, and facilities with the intended people at the intended production rate that is capable of meeting the customer's daily production rate (as specified on the Request for Quotation) and results in components that are correct to drawing and meet *all* customer's specifications. Typically, the volume produced will be sufficient to supply the customer with one full day's production.

On the other hand, the following list will give the reader an indication of what is not a complete PSW. This list is not an exhaustive one, but it gives the reader a direction that should be followed. The list is:

- Parts produced at the facility supplier and not at the production source.
- Parts made from 1 impression tool, but production intent is from 4 impression tools.
- Parts produced meet the ramp-up volume but not the full laid-down volume.
- Parts are produced off one line, but in production, three lines are required (*e.g.*, wiring looms).

To avoid issues of the PSW, we recommend:

- PSW runs should be conducted from a run of a minimum of one day's production (a day's production is identified from the Request for Quotation), run at the intended production rate. Any exceptions require an Alert signed by customer's engineering, Purchasing and Manufacturing.
- Any part not meeting the required production rate (specified on the Request for Quotation) is not PSW and requires an Alert signed by customer's engineering, Purchasing and Manufacturing.

- Any part that does not meet any of the quality or specification criteria is not PSW and requires an Alert signed by customer's engineering, Purchasing and Manufacturing.
- Any part not meeting the statement for "what is" a good PSW is not a PSW and requires an Alert signed by customer's engineering, Purchasing and Manufacturing.
- At a Program Review, whenever PSW data is shown, it should always state % PSW and separately % PSW (interim) with an approved alert.

In some cases, the supplier may be issued a conditional PSW *only and only if* everything is acceptable except the results of some testing that have been conducted but for some reason are not reported yet. Obviously, when the results are available, the final determination will be made.

In addition, some customers demand that part of the PSW must include the evaluation of capacity volume. If that is the case, the volume produced for the PSW run should be identified on the PSW. In all cases, the PSW is always a *legal document.*

CAPACITY ANALYSIS

Overview

One of the areas that has been overlooked or misunderstood by suppliers is the issue of capacity. In fact, capacity problems are a concern about 30% of the time as production, delivery, and volume fall short of the expected performance. The simple definition of capacity is a study that regularly is being conducted to get confirmation of supplier capacity to support new model launch and/or manufacturing plan changes. Supplier capacity means that the supplier is capable of providing enough product to sustain the customer's requirements. It is performed under the following conditions. However, the reader must understand that the timetable given here is approximate and may vary depending on the product and or customer requirements:

- New model confirms capacity studies are conducted at 2 weeks post final data judgment milestone at 17 months before Job 1. (Obviously, the time is organizational and product dependent.)
- Pre-Launch Readiness capacity studies (if appropriate and needed) are also conducted at 7.5 months for new model programs. This should align with tool trial (TT) build milestone.
- Capacity studies are conducted for planned capacity uplifts both within the new model launch timeline and post-launch.
- Capacity studies are also conducted to evaluate alternative product offerings to customers (mix rate changes).
- There is no difference in supplier response expectations for both "What If" and confirm studies.

Goal of Capacity

In no uncertain terms, capacity must be understood as a process, NOT a form. As such, data must be used to confirm that a supplier can produce the required *volume* of *quality parts* in the planned *operating pattern* at the planned *production rate*. The process starts with:

- *Planning*: Planning and asking whether or not the planned manufacturing equipment and processes projected to be capable of producing the required volume during planned operation patterns are accurate?
- *Verification*: Do the demonstrated run at rate and capacity verification stages of the PPAP results validate that the equipment and process can produce the required volume at the appropriate efficiency?
- *Managing growth and shortages*: Can the equipment and process supply additional volume? What can be done to increase production and efficiency? (The supplier must have at least a 10% buffer capacity beyond the requirement.)
- *Communicating*: Do suppliers cascade capacity study requirements and commitments to their affected plants? (Quite often the requirements are assumed to be cascaded but in reality they are not – especially to the sub-suppliers.)
- *Identifying wrong assumptions of past performance:* Are suppliers misinterpreting new study as superseding the prior study? (In case of capacity shortages, communication with the customer must be immediate, clear, and with an action plan to be back on schedule. This communication cannot be emphasized enough because any failure in capacity volume will have serious consequences for customer's production requirements. It is imperative, then, that the communication is understood as crucial to mitigating supply risks to the plants. The earlier the notification of pending issues, concerns, or problems, the higher the likelihood of being averted.)
- *Identifying the process of when and how to escalate:* Is there an escalation process for reporting any type of deviation? If there is no appropriate action taken, escalate issues impacting supply to all concerned as soon as possible.
- *Recognizing wrong capacity estimates*: Are the capacity estimates appropriately calculated? Sometimes, capacity reporting and data discrepancies are the causes for errors. The result will be an overstated or understated capacity reported to the customer and of course the data will not be aligned with the capacity verification. Here, we must note that the most common issue is the overstated capacity. The reason for this is more often than not, the appropriate allocation has not been accounted for or the correct OEE has not been recorded.
- *Identifying insufficient information from customer*: Are there sufficient information for appropriate decision? If not, the supplier capacity uplift execution falls apart. Typical issues here are: delay, and tooling investment until uplift volumes are cascaded in customer releases. This is a major

concern, especially when these delays are close to design freeze. Sometimes this lack of information falls on the supplier itself by not communicating early enough about potential issues via an SREA. Very important to note here that any supplier process and or location changes to support capacity uplifts must be communicated to the customer.

- *Reviewing the meeting capacity requirements*: Are reviews conducted appropriately? This requirement is based on 5 days and the maximum production weekly requirements in 6 days. This means that the supplier is to manage its tooling, equipment, and facilities such that, during a 7-calendar day week the Average Production Weekly (APW) capacity requirements are to be met by operating the tooling, equipment and facilities based on a 5-day work week (unless otherwise dictated by government regulations. In some countries the normal work week is defined as 6 days). On the other hand, the Maximum Production Weekly (MPW) capacity requirements are to be met by operating the tooling, equipment, and facilities based on a 6-day work week. The remaining time during the week is reserved for completing the required tooling, equipment, and facility maintenance. If the Supplier is unable to meet the Average Production Weekly based on a 5-day work week, or the Maximum Production Weekly based on a 6-day work week, the supplier must contact their customer to develop a resolution plan to meet the capacity requirements.

Any exceptions to these requirements must be requested by the supplier and concurred in writing by the customer.

Environment for Capacity Planning

To validate a supplier's capacity plan, some type of capacity analysis report (CAR) form must be filled. [Do not confuse this report with the corrective action report, which is also designated as a CAR.] There is no standard form for all customers. However, it is imperative that both customer and supplier agree on a CAR, and it should be utilized to make a comparison between the supplier's planned manufacturing plan (Required Overall Equipment Effectiveness (OEE), for planned volumes) and their historical manufacturing performance. Relevant information on the CAR during capacity planning is and can be recorded in a tab form:

- Capacity planning
- Historical manufacturing performance
- Shared loading plan
- Supplier declarations, including sub-tier suppliers

The integrity of each tab is critical in completing a proper analysis – the selection of appropriate processes and parts for historical manufacturing performance, and the inclusion of appropriate changeover times (if applicable) is required.

The intent of the capacity planning analysis is to provide validation of the supplier's capacity plan. One of the very basic questions is: Does the supplier plan have an appropriate work pattern, with appropriate equipment cycle times, to meet the expected program volume adjusted for their historical manufacturing efficiency? Note that if historical data is not available, surrogate production data from a similar manufacturing process may be used.

To make sure that the basic question is answered, the supplier conducts the initial run at rate. This is a very important step because parts produced from a production stream (from a minimum of one production tool, line, process stream) at production feeds and speeds will give an indication whether the supplier can indeed produce what is expected. This initial stage provides an early indicator if supplier can meet future timing of delivery and production requirements.

In conjunction with the requirement to satisfy the appropriate requirements for producing parts at designed cycle times and achieving print specification requirements, a CAR form is required at run at rate to validate the supplier's ability to achieve all the Run-at-Rate requirements. This means typically a utilization of a short duration of production (~300 pieces although according to AIAG (2009 p. 3), this can be changed with the approval of the customer representative), it is also strongly recommended that a capacity planning document with historical manufacturing performance data also be reviewed – if it has not previously been completed. It also allows for the analysis of additional production streams as required for the completion of the quality verification. By adjusting the planned net ideal cycle time to account for additional production streams, the overall capacity can be analyzed in support of the quality verification requirements.

The second issue of planning is the concern for quality verification. Here, parts and processes from a minimum of one production stream (tool, line, facility, *etc.*) are evaluated for quality performance. Typical issues are: Dimensional, lab, and engineering specifications (ES) testing complete for this production stream (less appearance approvals). If the supplier is ready, this stage may be combined with the run at rate.

Production verification is the third stage of planning. Here, the complete actual stream production stream (tool, line, facilities personnel, *etc.*) intended for this specific program/launch is in place and operational. Dimensional, lab, and ES testing complete from all tools, cavities, molds, and production streams (including all appearance approvals). Supplier submits PSW. Quite often, this stage is combined with the second stage. It is only necessary if the supplier has multiple streams.

The last stage of planning is "the" capacity verification. Here, parts are produced from the complete actual production stream (tooling, equipment, facilities personnel). Capacity verification is demonstrated by yielding quality parts to meet a minimum or one day of the customer's production requirement (daily planning volume (DPV). If three shifts are planned, the capacity verification study *must* reflect that pattern of production.

In conjunction with the requirement to satisfy the mass production (full production) requirements for producing parts at designed cycle times and achieving print specification requirements, a CAR form is required at capacity verification stage to validate the supplier's ability to contain total program capacity. Again, as the capacity verification continues so does the Run-at-Rate event continues to be a relatively short duration of production (~24 hours), it is also strongly recommended that a capacity planning document with historical manufacturing performance data also be reviewed – if it has not previously been completed. At this stage, it is also assumed that all production equipment, tooling, personnel, and gauging are in place and that all production streams *can achieve* the required production rate.

OVERALL EQUIPMENT EFFECTIVENESS (OEE)

Overview

One of the most underrated and misunderstood components of the capacity study is the OEE analysis. OEE combines key manufacturing metrics to state the overall health of the production process. It can be communicated as a minimum required OEE or an effective/demonstrated OEE. The relationship between these two determines the feasibility of the manufacturing process to meet customer volume requirements. For a very detailed analysis of OEE, the reader is encouraged to see Stamatis (2010).

- *(Minimum) Required OEE:* It is the minimum OEE a supplier must achieve to support the customer volume requirements. Warning: the traditional 85% initial OEE may not be accurate and may not be as good as one thinks. The best OEE is 100% anything else is an opportunity for improvement. If there is more than 100% the cycle time is miscalculated. For a detailed discussion on this, see Stamatis (2010).
- *Demonstrated OEE:* This is the OEE a supplier demonstrates – through a PPAP event, or through historical/surrogate analysis.

Both types of OEE depend on the following three calculations:

Availability: Includes operating pattern, downtime, and allocation percentage. The actual formula is:

$$\text{Availability} = (\text{Operating time}) / (\text{Net available time})$$

Performance efficiency: Rate at which parts are produced compared to the ideal rate. The actual formula is:

$$\text{Performance} = (\text{Total parts run}) \times (\text{Net ideal cycle time}) / (\text{Operating time})$$

Quality rate: Comparison of *good parts* to *all parts* produced. The actual formula is:

$$\text{Quality rate} = (\text{Total parts run}) - (\text{Total defects}) / (\text{Total parts run})$$

The relationship between these three components produces the minimum level of efficiency required to meet customer volume requirements, or required OEE.

Using OEE as the unit of measure also allows assessment of a supplier's capacity for future production using current production performance data, surrogate OEE, or historical manufacturing performance.

Comparing the required OEE and demonstrated OEE determines the supplier's capacity risk.

The reader hopefully has noticed that the OEE combines key manufacturing metrics to state the overall health of the production process. It can be communicated as either a minimum required OEE or an effective or demonstrated OEE. The relationship between these two determines the feasibility of the manufacturing process to meet customer volume requirements. This relationship is shown in the actual form of the OEE with the appropriate headings and calculations of: (a) Net Ideal Cycle Time, (b) Total Good Parts Produced, and (c) Net Available Time (Stamatis, 2010). Remember that these are the same inputs used on the historical manufacturing performance sheet.

Evaluating the OEE

Once the calculations have been completed, then an analysis of both the data and results is appropriate. One of the most important questions that must be asked is: Is demonstrated (surrogate) OEE greater than or equal to (≥) to required OEE? The answer to this question will be used to determine the capacity risk at all the appropriate milestones and the final delivery date. When evaluating the OEE, one should remember that the demonstrated OEE is based on: (a) Surrogate data in capacity planning, and (b) actual run data during run at rate and production verification stages.

On the other hand, the required OEE is the minimum level of efficiency required to meet customer demand, based on: (a) Customer's capacity requirements, and (b) supplier's manufacturing plan.

If all processes have a favorable answer to the above question, there is no risk to capacity. Focus on continuous improvement and transferring lessons learned to new model processes. On the other hand, if *any* process has an *unfavorable* answer to the above question (*i.e.,* Demonstrated OEE < Required OEE), there is a risk to capacity. The capacity gap between demonstrated OEE and Required OEE must be closed to meet customer volume requirements. Obviously, the supplier if there is a deficiency in capacity has two possible alternatives. The first is to increase the Demonstrated OEE, and the second to decrease the Required OEE. The best option is to work on improving the Demonstrated OEE. [There is no threshold for the initial OEE. Whatever it is, make sure it is the right one for it serves as the benchmark for the future OEE. That means as the supplier progresses, the OEE should get better. If the initial OEE is below 85% try to identify what are the reasons and correct them.]

Identifying potential capacity risks early in the timeline cycle allows necessary time to implement improvements to minimize the risk to launch. Regardless of *when* the capacity gap is identified, the process of resolving the capacity gap is the same. This is true even when capacity gaps are identified after launch.

Some Specific Actions for Improving OEE

Improving Demonstrated OEE: By improving surrogate processes and duplicating lessons learned, capacity gaps can be eliminated, and the overall "health" of the supplier is improved. Remember that the Demonstrated OEE is the actual level of efficiency achieved by a supplier in historical or current production. For Capacity Planning, surrogate processes are used for Demonstrated OEE; at run at rate and capacity verification, actual performance data is used. So, in Capacity Planning (forward-model), the Demonstrated OEE is made of historically similar (surrogate) processes. At run rate, the Demonstrated OEE is made of actual production data. This is always true after this stage as well, whether before launch or in a capacity constraint situation. The approach to improving Demonstrated OEE – whether Surrogate or Actual – is the same.

Some Specific Improvements on the Components of OEE

For Availability:

- Track and Pareto losses to reduce unplanned downtime.
- Identify special and common causes.
- Analyze Mean Time Between Failure (MTBF), Mean Time To Repair (MTTR), *etc.*
- Utilize Rapid-Response Maintenance Teams and Hourly Count Production Boards.
- Prioritize equipment for maintenance activities.

For Performance Efficiency:

- Ensure performance to constraint ideal cycle time.
- Identify blocked and starved conditions; ensure cycle times support Takt time requirements. Make sure you distinguish between Tact (theoretical) and Takt (actual) time.
- Machine/Equipment – Have an engineering study to identify the gap, and assignment of appropriate technical resources for closure.
- Human Resource – Address the shortage of operators, imbalanced work, or lack of standardization.

For Quality Rate:

- Track, Pareto, and reduce scrap and rework losses.
- Install poka yoke/error proofing devices.
- Communicate good vs. bad parts (include boundary samples).
- Allocate problem-solving resources and process experts to launch.
- If the capacity analysis is performed early in the timing process, there is time to implement OEE.

Demonstrated OEE

Demonstrated OEE is generated from actual performance data for each process and is used to assess risk when compared to the Required OEE for each process. At the run at rate and capacity verification stages the Demonstrated OEE is generated from the performance data of the run event. Generally, there are no significant changes from the current capacity standard at both stages relative to the inputs required completing the analysis. However, by moving the initial assessment of capacity earlier to say program approval (about 30 months before Job#1), actual performance data may not be available during capacity planning. As such, the Demonstrated OEE is generated from performance data of *surrogate* processes.

When selecting a surrogate process, the best surrogate should be aligned with the planned process for the new part. Considerations include:

- Part complexity
- Technology used in the production
- Manual or automated process
- Greenfield (new) or Brownfield (existing) site
- Process layout comparison
- Volume, operating pattern, *etc.*
- Similar part size and cycle time
- Potential cultural differences
- Surrogate processes selected do not necessarily need to be from a common part or time (*e.g.*, milling processes, welding processes). On the other hand, it is important to remember that when selecting surrogate processes, it is not required that the overall parts are exact. The goal is selecting manufacturing processes that are similar to the new part's manufacturing processes.

To obtain historical OEE performance data for each process, we need three known data inputs. They are:

1. *Net ideal cycle time* (NICT): NICT should be verified through observations or design, ***not*** engineering standards. NICT must *not* include any "baked-in" efficiency losses. NICT is the best, sustainable cycle time achieved in production (steady state, not ramp up).
2. *Net available time* (NAT): NAT is the amount of time that the process was in production for the specific part number. NAT also *excludes* planned downtime (lunches, breaks, *etc.*) but *includes* unplanned downtime. This means that if the process was shut down early on Friday because requirements were met do not include the time that the process was shut down. For shared processes, NAT must also include changeover time.
3. *Good parts produced* (GPP): GPP includes only the good parts through the process – no scrap or rework parts (first time through).

Enablers to Improve the Demonstrated OEE

- With each action, the Demonstrated OEE should be monitored to validate the OEE improvement.
- Improving the Demonstrated OEE must be the first approach considered to close capacity gaps and eliminate risk.

Other enablers:

- Install strategic buffers to prevent starve conditions.
- Establish energy room to communicate metrics and progress to plan.
- Run simulations with multiple iterations – changing input variables (cycle times, downtime, quality rate, buffer sizes, MTBF and MTTR, *etc.*) to optimize throughput.

Analyze results and implement actions:

- Ensure cascade of lessons learned (TGW and TGR) from previous launches.
- For shared equipment, reduce changeover times.
- Review/update preventative maintenance schedules.
- Investigate incremental tooling and equipment potential changes.
- Ensure there is a method of verifying the results. Translate the results to OEE so that a common language is being used to track improvements.
- Make sure that action items are prioritized appropriately.

Required OEE

Required OEE is directly proportional to the required good parts. However, if the consideration of other losses (scrap, rework, changeover, *etc.*) exceeds the maximum possible parts, the manufacturing plan is not feasible. This condition can occur *even when* the required OEE appears reasonable. Remember, Required OEE is the *minimum level of efficiency* that is required to support the demand, based on the defined operating pattern, allocation percentage, and operating parameters. Once all potential losses are assessed, a process may not be feasible, even if the Required OEE < 100%.

All *value streams* are required to be assessed for capacity risk. That is, all processes that are utilized for production require assessment for risk. This may require the use of more than one document, and is consistent with PPAP requirements. For end item part numbers that have multiple *value streams*, the Capacity Planning Analysis must begin by looking at a single *value stream*. This is due to considerations made to individual process scrap losses, which have impact on the individual process Required OEEs. In effect, the Capacity Analysis is generally set up on a ***part-specific*** basis. However, depending on the part and manufacturing process, it *may* be appropriate to group families of parts, or to group several manufacturing lines. If in doubt, check with the customer for approval.

Improving Required OEE

By adding operating time, reducing up-line scrap loss, or decreasing net ideal cycle time, capacity gaps can be eliminated. The minimum level of efficiency required by a supplier to meet customer's volume requirements, based on the supplier's planned operating pattern and manufacturing assumptions. The Required OEE for each process must be understood and identified as early as possible by both customer and supplier. For each process, the Required OEE is the *target* – capacity gaps can only be identified by comparing Demonstrated OEE to this target. The basic question here is: Are the Required OEEs for each process feasible? Specific items with reducing the Required OEE with minimal investment that may be considered are to expand operating pattern, add extra shifts of expanded hours, and consider overtime. Another option is to reduce constraint cycle times by: (a) Machine/Equipment Design, (b) Coordination with equipment suppliers to reduce cycle time, (c) Introduce faster processes when and if possible, and (d) Track and optimize cycle times at ramp-up.

Shared Loading Analysis

Shared loaded analysis is the evaluation of how much time is allotted for production using a stream for the customer. Make sure that the allotted time is rated correctly and fulfils the requirements of the customer. Generally, there is only one exception to this rule where total allocation is calculated. The exception is for machining and outsourced e-coat. To complete a "Shared Loading Plan" for a process, the following information is needed:

1. Average Production Weekly (APW) and Maximum Production Weekly (MPW) volume information adjusted for downstream scrap losses, along with the NICT for all parts that are planned for production on the specific process; average demonstrated OEE from the most recent production performance data. The "Shared Loading Plan" validates the accuracy of the planned allocation percentage for the process, adjusted for the Demonstrated OEE.
2. The equipment/process is not oversold considering the total book of business of the process.

Total Allocation

For machining, if the total % allocation is 54.2% for both APW and MPW, this means that, adjusted for the supplier's demonstrated OEE, the operation can contain all business (customer and non-customer) without exceeding the acceptable operating patterns. However, for Outsourced E-Coat, if the total % allocation is *greater than 100%,* the process is oversold, *i.e., capacity is at risk* (must follow up with action plan). The Total % allocation can be addressed in the same ways that gaps are addressed between Required OEE and Demonstrated OEE – by utilizing capacity gap closure to improve Demonstrated OEE, by reducing Required OEE (and thus, the part-specific allocation percentage), or by offloading some of the business.

Reaction and Plan Expectations

When all the data have been collected and the analysis has been completed, it is time to evaluate the results. This evaluation is based on (a) reaction to the data, and (b) planning for the expectations. Both are dependent on:

1. An assessment of current capacity and potential supply impact by:
   a. Understanding the maximum possible output.
   b. Examining alternate manufacturing options.
   c. Establishing output tracking of constraint operations.
   d. Reviewing the supply plan based on customer release requirements as far in the future as it is available.
   e. Establishing supply mitigation if there is supply shortage projected.
   f. Establishing contact with appropriate customer representative if there is permanent degradation in output.
2. Communicating with customer immediately. This means contact the appropriate customer representative immediately and convey any issue, root cause, and supply status.
   a. Presenting a supply protection plan.
   b. Executing the shipping and actions as agreed.
3. Corrective action. Here, the supplier is required to investigate process and system escape contributors to root cause(s) by utilizing formal problem-solving techniques. Furthermore, the supplier must:
   a. Establish permanent and systemic corrective actions.
   b. Update preventive maintenance plans and/or process control plans with permanent corrective actions as applicable.
   c. Capture systemic improvement actions in customer's manufacturing site assessment system.
   d. Implement corrective actions per plan and validate results.

Capacity Information Maintenance

As in any system so does the capacity has to be maintained on a regular basis. Typical appropriate maintenance approaches are:

1. At a minimum, suppliers are required to update their capacity information in the customer's capacity planning systems on a quarterly basis – or as required by the customer.
2. Suppliers must also update their capacity information in the customer's system (if available) as soon as there is a significant change in their operating output such that you are not able to support previously declared capacity in the customer's system. For significant drop in capacity, please notify the customer for immediate supply risk mitigation.
3. Besides maintaining capacity information in the customer's system, suppliers must monitor all customer's releases for significant amount of time (in some cases for 6-months releases) and communicate any supply constraints to the appropriate customer representatives.

Some specific items of reviewing for appropriate maintenance are in the areas of:

1. Study:
    a. Confusion and misinterpretation due to multiple studies.
    b. Supplier response errors.
    c. APW/MPW operational practices *not* consistent across supply base.
    d. No sub-tier confirmation.
2. Tool order:
    a. Missed tool funding due to supplier response errors.
    b. Supplier execution:
        i. Delayed tooling kick-off through supply chain.
        ii. Inadequate resources impacting timely execution.
        iii. Poor sub-tier planning and execution.
        iv. Does *not* believe increased releases are real.
3. Capacity verification:
    a. Overcommitted and overstated capacities.
    b. Sub-tier *not* verified prior to capacity confirmation with customer.
    c. Changeover times *not* accounted in shared lines planning.
4. Sustainable manufacturing:
    a. Domino effect from the lack of quality of event upstream.
    b. Lack of ability to recover from non-robust plans.
    c. Unplanned downtimes due to extended operating patterns (lack of preventive maintenance).

Thought Starters to Improve Supplier Capacity Volume

OEE improvement actions and maximize the following: (a) Equipment availability, (b) performance efficiency, and (c) quality rate. Specific questions may be:

1. Are there hours, days, and shifts available so that the supplier can run extra parts during the week? Understand the operating patterns.
2. Is there any opportunity to improve the cycle of time constraint station or stations? Develop Performance to Takt Tools (Rational: Performance efficiency improvement).
3. Does the supplier collect accurate data for downtime and scrap/rework?
4. What are the major downtime issues? Deep diving into the top (5) downtime issues. Are there any available Pareto charts on downtime for (4) weeks? (Rational: Equipment Availability Improvement).
5. What are the major Scrap/Rework issues? Deep diving into the top (5) scrap/rework issues. Are there any available Pareto charts on downtime for (4) weeks? Parts lost to rework/scrap (Rational: Quality rate/performance efficiency improvement).
6. Can the changeover process timing be improved? Review the changeover process at supplier site. Apply SMED opportunity, lean tools, and principles (Rational: Quality rate/performance efficiency improvement).

7. Are there any opportunities to improve the consumable change/over (C/O) timing?
8. Is the supplier doing too many changeovers?
9. Are there any opportunities to implement PM scheduling, cleaning during the changeover, or breaks/lunches? (Parallel Tasks).
10. Are there any opportunities to run the rework/reject parts during the breaks/ lunches at the final inspection cell?
11. Are there any machine start-up checklists available? Can the start-up checklist to be performed during lunches/breaks?
12. Does the supplier have a complete spare part available (electrical/tooling, *etc*.)?
13. Does the supplier have adequate raw materials on the site for the increased production volume? (Especially parts that are shipped from Europe or Asia.)
14. Are there any opportunities to implement the tag relief during lunch and breaks?
15. Are there any stations or machines in bypass mode that impacts the efficiency throughput?
16. Does the shift change occur in a timely manner? Minimum impact on the equipment availability.
17. Is there any opportunity to optimizing the start-up between each shift? Minimize/optimize production meeting between the shifts, *etc*.
18. Are the production counts shifting to a large fluctuation? Deep dive into the root cause analysis and standardize operation procedures.
19. Is there any opportunity to add stations or stations to the constraint station/cell to sustain volume increase w/a minimum capital investment? Is there any opportunity to make more parts in the offline station? Is there any opportunity to outsource the parts?
20. What are the next constraint stations or cells beside the current cell?
21. Is there any opportunity to develop a short-term/long-term road map to capture the improvement job per hour/Average production weekly (JPH/APW) Pareto to sustain the customer production demand?

MSA

Overview

There is a truism that all of us should keep in mind. That is: Good outgoing quality depends on good incoming quality. The incoming quality for any customer depends on their supplier's quality. Therefore, the supplier's quality is an issue of both strategy and execution. Strategy is fundamental because it will define the necessary needs for excellence, market intelligence, good timing policies, risk management, and transparent collaboration of customer-supplier relationships. Execution, on the other hand, is critical in understanding and implementing (or creating in some cases) the analytics of the appropriate and applicable *data* that have been defined and collected.

Between strategy and execution, the supplier has the responsibility to focus on the "real data" of the organization with the intent to either modify or change drastically the way they are doing business for improved or continual improvement. This desire of being better has been accentuated with the latest recalls in the automotive industry. In fact, Walsh (2014, p. 11) reports that the "GM recall scandal has manufacturers scrambling to review documents, practices to protect against potential liability."

We all have heard the old saying of "what is measured it gets done." So, for the past 20 years the Quality Profession has focused on the issue of data. If it is worth measuring then appropriate and applicable data must be available, retrievable, and analyzed. However, to have data, it is not a difficult thing. What is difficult is to have consistent and comparable data. This is of profound necessity because unless the data of what you are counting as "real" is real, the answer you will get is at best "not sure." In the case of the supplier, measurements are a key item of concern in many of their processes and operations. So where do we start? The following may help:

1. Start with a simple unit of measurement you can use as a basis for rationalizing all your data. This can be optimized by thinking in terms of what the customer wants. Next, coordinate the process steps so that they follow a clear sequence tied to this unit of measure in real time. Finally, if possible, automate the process so that the measurements are consistent.
2. Keep it relevant. All data and measurements should focus on identifying key characteristics as to where they are (baseline) and in what direction they are moving (monitoring).
3. Automate wherever possible. This will provide real-time measurements of real data, but in addition to the increased speed, it will provide accurate, consistent, and reliable data for your analysis.
4. Make sure the instruments (methods) used are accurate, reliable, and repeatable. Remember that the instrument that the measurement is being taken must be at least 10X more sensitive than the measurement recorded.

In essence, data and measurement must contribute to some form of decision to fix or improve a process. From a supplier's perspective, the following items may be considered as possible sources for data:

- *System complexity*: The use of multiple suppliers in multiple locations and attendant inventory storage and transportation waste. Examples of measurement may be the cost of the system or delays created by complexity.
- *Process efficiency*: Optimizing process characteristics to control quality. Examples of measurement may be key input and output characteristics, especially for special characteristics that the customer and supplier have identified as important to be controlled and monitored.
- *Lead time*: Generated by purchasing when negotiating lead time with the supplier or delays built into the transit process. Examples of measurement may be the cost to the process or system for delays and waste of excess inventory or processing.
- *Transportation*: Wasted effort to ship the product or wasted distance in the transportation process. Examples of measurement may be the cost of transportation, especially premium freight.

- *Space (facilities):* The space needed to transport or store product prior to use. Examples of measurement may be a facility (or cell) of square footage and the cost associated with that space.
- *Inventory*: Inventory beyond what is needed to serve customers and satisfy the process. Examples of measurement may be the cost of inventory and the cost of maintaining it.
- *Human effort*: Wasted movement and motion of an operator(s), or losses due to accidents. Examples of measurement may be wasted time and the operator's compensation insurance costs, as well as effort for rework.
- *Packaging:* The costs associated with over- or under-packaging resulting in waste or product damage in transit. Examples of measurement may be the cost of premium freight, repeat shipping, and or product replacement.

After reviewing these areas then specific considerations for quality improvements should be considered with the following characteristics in mind:

- Is it the right product? If not, how can we improve it?
- Is it the right quantity? If not, what is missing? Can we make it right?
- Is the right service available? Is the right service provided? If not, what can we do to make sure that the appropriate service is applied in satisfying the customer?
- Is the right condition to collect data? If not, what can we do to make sure that the conditions are appropriate and applicable for appropriate data selection?
- Is the right place and time to collect data? If not, what are the conditions to make sure that the place and time are appropriate and applicable for collection of the data?
- Is the frequency and source appropriate for representative data to be collected? If not, what can we do to mend this situation?
- Is the cost (price) right for what we are doing? If not, what did we miss? How can we recover data to reflect the right cost or price?
- Is safety an issue? If yes, what can we do to minimize the risk of accidents?
- Is the data taken from a confide space? If yes, can we arrange the data to be drawn from alternative source or an alternative method?
- Are the operators trained in sampling and measurement techniques? If not, what would it take for them to be appropriately trained?

To be sure, these questions are not one-to-one relationships with the items that we mentioned above. However, they are important and they should be viewed and investigated because they could very well be seen as root causes of poor performance as long as the "right" data are collected.

What to Look for:

Repeatability Issues (Within Operator)

No gaging instructions or operational definitions

- Lack of skill in reading vernier scales.
- Clamping sequence differs from trial-to-trial.

- Loose gage features/gage not rigid enough.
- Measuring location on part differs trial-to-trial.
- Semi-destructive testing changes the part (squeezing soft parts).
- Not enough gage resolution.
- Rounding away (up and down) gage resolution.
- Instrument needs maintenance.

Reproducibility Issues (Between Operators)

- No gaging instructions or operational definitions.
- Differences in skill in reading vernier scales.
- Clamping sequence differs from operator-to-operator.
- Different (operator-owned?) measurement devices.
- Ergonomics: Operator size, height, strength, *etc.*
- Different rounding methods, operator-to-operator.
- Temperature of the gage over time, *i.e.,* expansion.

Key to Successful MSA

- Define and validate measurement process.
- Identify known elements of the measurement process (operators, gages, SOP, setup, *etc.*).
- Clarify purpose and strategy for evaluation.
- Set acceptance criteria – Operational definitions.
- Implement preventive/corrective action procedures.
- Establish ongoing assessment criteria and schedules.

Possible Causes for Bias

- Instrument needs calibration
- Worn instrument, equipment, or fixture
- Worn or damaged master, error in master
- Improper calibration or use of the setting master
- Poor quality instrument – design or conformance
- Linearity error
- Wrong gage for the application
- Different measurement method – setup, loading, clamping, technique
- Measuring the wrong characteristic
- Distortion (gage or part)
- Environment – temperature, humidity, vibration, cleanliness
- Violation of an assumption, error in an applied constant
- Application – part size, position, fatigue, observed error (readability, parallax)

Possible Issues with Linearity

- Instrument needs calibration, reduce the calibration interval
- Worn instrument, equipment, or fixture

- Poor maintenance – air, power, hydraulic, filters, corrosion, rust, cleanliness
- Worn or damaged master(s), error in master(s) – minimum/maximum

One of the benefits of using Minitab is the fact that it has one of the best "help menu." The person who is using Minitab at any time may consult the help menu for either help in selecting a particular test or help in the interpretation.

11 Available Software for Customer Service

OVERVIEW

Our world today is focusing on speed. As such, the advent of software programs in the area of customer service may be useful to expedite problem solving as well as innovation in the area of customer satisfaction. After all, the focus has been and will be that serving customers is the goal of any organization. If not, it should be!

A good software should have at least the following abilities:

- Manage email, live chat, and social media requests from one centralized tool.
- Build a customized document.
- Use AI features to help get customers the answers they need quickly.
- Create customer profiles, so you are able to see past interactions.
- Track metrics that help the organization to be better at customer service. Some examples are: Help built reports so you can see how your team is performing from day one; Generate reports that show what times of day are busiest, common reasons for support requests, average response times, and other useful organizational information as needed.

Of course, using traditional tools is a good way to start out. However, the use of computers has also complicated the selection. Figuring out which customer service tool best serves the organization can be a tricky task. The selection is based on what is needed and whether it is flexible enough to meet future needs. Of course, these considerations are a function of the available budget.

As important as the budget is, we must always remember that if it is worth doing something, then it must be done appropriately and "done" right the first time. This is a warning, of course, because the selection of "the" appropriate and applicable software is indeed complicated and confusing for many. The implications of buying and using the wrong tool, having dissatisfied customers, and frustrated team members are far worse. So, take your time, be thoughtful, consider your options, and make the most informed decision you can.

Always remember that "no matter which software you choose," it is the service you deliver to your customers that matters. Do not fall for the "best." Do not let the search for the "perfect customer service software" stop you from defining and delivering the service experience that will keep those customers coming back. Remember also that many programs try to be a one-size-fits-all solution.

DOI: 10.1201/9781003650614-11

WHAT IS CUSTOMER SERVICE SOFTWARE?

Customer service software is a set of tools used to collect, organize, respond to, and report on customer support requests. A customer support system may be used to manage one or many communication channels, including email, live chat, messaging, and self-service, and it may also integrate with external communications tools like social media or group chat systems. Organizations typically use cloud-based customer service software to enable faster, more efficient customer support delivered by multiple customer service agents all working within the same tool. So one may wonder, what are the benefits of using a customer service platform? To be sure, one can certainly deliver great customer support without using specialist software, and many online businesses start out with nothing more than a free email account. Soon, though, growing companies tend to run into some limitations and rough edges. Using a specialized customer service system helps you to create a better customer experience that improves customer loyalty. The key ways you can use a specialized tool to enable those experiences are:

1. *Give more responsive and more consistent support*: Dedicated support software features like automated workflows, tagging, knowledge base integration, saved replies, AI, and more help your team spend more time helping customers and less time fighting their tools.
2. *Gather customer insights*: Identify, collect, and organize all the helpful feedback, feature requests, bug reports, and use cases so they can be used to improve customer satisfaction.
3. *Work better together*: Customer service software enables you to reduce duplicated work, keep track of customer queries, coordinate a response across multiple teams, and deliver up-to-date answers.
4. *Analyze and report*: Use the built-in reporting features of customer support software to understand changes in support volume, team productivity, type and size of customer, and much more.
5. *Scale up your service*: Maintain high-quality service as you grow by using tools that can coordinate multiple teams of agents to support an enormous customer base, using organizational features, automations, and third-party integrations. When you have more than a couple of people working together to support customers, using specialized customer service software is the right choice.

Now the real question becomes: What are the different types of customer service tools? Building out a customer service program is like building a chair. You could – in theory – build either one with just two or three tools, but the overall quality and efficiency of your efforts would be greatly impacted. Whether you are building a chair or a support team, there is a continuum of tools ranging from broad to specialized, and the tools you end up investing in, will be a reflection of the stage you're at in the craft and the resources available to you.

So, at this point, one may wonder as to what are the most common customer service software features? Obviously, the feature set of software platforms built for customer service covers a wide range, but it can be generally categorized into six major focus areas.

1. *Collecting*: Collecting features help you answer the question, "How do we get customer communications into this system so we can handle them?" They provide the first point of interaction between the customer and the customer service software. Common examples include a support-type email address, "contact us" forms that funnel messages into a support inbox, phone numbers, messaging systems, and APIs. [APIs are mechanisms that enable two software components to communicate with each other using a set of definitions and protocols. For example, the weather bureau's software system contains daily weather data. The weather app on your phone "talks" to this system via APIs and shows you daily weather updates on your phone.] In some cases, customer service software tools may include built-in interfaces for some channels and may integrate with external providers for others.
2. *Organizing*: Organizing features are for taking all that incoming communication and creating useful structures so customer service teams can manage high volumes, understand what needs to be done, and respond in a timely manner. Organizational features in customer service software cover both tools for manually arranging things and tools for acting automatically. Common examples include: Folders and views to easily look at subsets of the conversations; Tags to label conversations for handling and reporting on later; Workflows for taking actions automatically or manually, such as adding tags, assigning to the right person, or setting priorities; Multiple mailboxes to separate different sets of customers or communications from each other; Custom fields to capture useful information in a structured way about the request or the customer.
3. *Collaborating*: Collaboration features allow multiple people to effectively work together on the incoming support volume, from frontline support folks to subject experts and business operations folks. Common collaboration features in customer service software include: Teams so that each group in your organization can see the conversations most relevant to them; The ability to assign conversations so that a conversation that requires a particular person's skill can be clearly allocated to them; Internal notes to allow staff to pass information between team members while keeping it associated with the customer conversation; Private notes to give a team member a heads up about relevant information in a conversation without making them responsible for replying to the customer.
4. *Responding*: These features encompass all the ways a reply can be sent to the customer in question. Responding features can include: Text editors for composing and sending responses to individual emails, live chats, and messages; Social messaging tools to respond publicly to incoming requests; Knowledge base systems for creating and publishing help documents to share with your customers.
5. *Integrating*: Many customer service software tools offer direct integration with other systems as well as APIs for programmatic integration into even more places.

Integrations allow a company to connect its customer service data with tools like: Slack or other communication systems for keeping your team up to date; CRM software for additional customer history and context; Shopify and other ecommerce services you use to serve customers; Social tools for other forms of customer contact; Internal systems that inform customer service decisions.

6. *Analyzing and reporting*: Features for reporting and analysis in customer service software allow companies to better understand things like who their customers are, what they are trying to get done, where their customers run into trouble, and what they need. Common customer service software reporting features cover things like: Time to first response; Customer satisfaction levels; Time to resolution; Incoming request volume over time; Common request categories.

Very important to note here is that some of the features above are common across nearly every customer support platform; others are less common or are implemented quite differently. To help you find acceptable software for your customer support team, we have identified some software in the pool of many that you may consider. The software that we identify here are a suggestion. However, most of them are quite practical, flexible, easy to use, and most importantly, provide the user with the ability to grow with time. A list of several software is provided based on the: https://www.helpscout.com/playlists/customer-service-software/. (For more detailed information on these software packages, see Appendix A.)

1. Zendesk: Good if you have large teams.
2. Aircall: Good if you use call center support.
3. Freshdesk: Good for enterprise teams.
4. Olark: Good customer service software for stand-alone live chat.
5. HelpDocs: Good for stand-alone knowledge bases.
6. Help Scout: Good for providing a great experience for both your customers and your support team – quite easy to use.
7. Text-Em-All: Good for automating phone communications.
8. Buffer: Good for social media management.
9. Helpshift: Good for providing support in a mobile app.
10. HubSpot Service Hub: Good for large businesses that already use HubSpot.
11. Salesforce Service Cloud: Good for teams providing support via Slack.
12. LiveAgent: Good for teams interested in gamification.
13. ServiceNow: Good for IT teams at enterprise companies.
14. Zoho Desk: Good for sales-focused teams.
15. Jira Service Management: Good for IT teams.
16. Sprout Social: Good for enterprise companies.

To determine which tool is right for your organization, consider the following nine types of customer support software.

1. *Help desk software*: Help desk software is typically one of the first tools teams invest in. It helps centralize all your service-related customer interactions, and, depending on the tool, it can handle support requests

from multiple channels like email, live chat, social media, and phone. At its core, help desk software lets you manage and streamline customer conversations to create a better customer experience and agent experience.

2. *Shared inbox software*: Shared inbox software is an email tool that allows multiple people to access and respond to messages sent to a specific email address. There are generally also other organizations and automation features included to help effectively manage customer conversations. Here, one may wonder! But isn't that the same thing as a help desk? Yes. Sort of. Shared inbox software is like a lite version of help desk software, since it tends to focus mainly on email interactions and not on the additional channels that a help desk may cover.
3. *Phone support software*: Presently, phone support (also referred to as "voice") may seem like a relic of the past, so it may surprise you to know that over 50% of customers use the phone to contact customer service. Furthermore, it is not limited to one demographic: People of every age, gender, and background opt for phone support. Therefore, no matter what industry you are in, you probably have customers who would be happy to give you a call. Phone support and contact center software is a more modern approaches to handling those phone-based interactions.
4. *Knowledge base software:* Knowledge base software is a tool that allows you to create, store, organize, manage, and share self-service content with an audience. Things like FAQ (Frequent Asked Questions) pages, video tutorials, and how-to articles are all common types of content housed in a knowledge base. Generally, there are two main types of knowledge bases: internal and external. With an internal knowledge base, content is private and may require the user to log in to access the information, whereas an external knowledge base is public and accessible to anyone. It may be helpful to think of an internal knowledge base as geared toward your employees, while an external knowledge base is geared toward your customers. Some tools focus more on one use case over another, but there are also some capable of doing both well.
5. *Live chat support software*: Sometimes, customers want to talk to a live person. Maybe they are not finding what they need in your help documentation, or they want to talk through an issue in real time, and email simply does not do it. In those instances, live chat is a great option since it offers the immediacy of phone support while being less resource-intensive. It also has a much higher average customer satisfaction rating when compared to phone – 82% satisfied for live chat vs. 44% for phone. (https://www.superoffice.com/blog/live-chat-statistics/). In order to offer live chat support, you first need access to live chat software. Like some other tools on this list, live chat software can be a stand-alone product, but it can also be included as part of a help desk's larger suite of tools.
6. *Messaging tools*: Messaging tools are a broad category referring to software that allow you to do some sort of proactive support. That could come in the form of chatbot software, proactive messaging software,

or some combination of the two. In any case, the core goal of a messaging tool is to reduce friction in some way or another for the customer. Data confirm people generally prefer tasks that require less effort (https://cognitiveresearchjournal.springeropen.com/articles/10.1186/s41235-020-00226-5).

7. *CRM software*: Customer relationship management (CRM) software is a tool that lets businesses manage and track customer interactions. It is most commonly associated with sales, but there are some customer success applications for it, too. The primary way it can help with customer success is through personalization. With the data you have collected, you might be able to better anticipate needs and curate a specific experience for each customer based on where they are in their customer journey. Research shows personalization pays off: One study found that 80% of consumers are more likely to purchase from a brand that offers a personalized experience (https://www.forbes.com/sites/blakemorgan/2020/02/18/50-stats-showing-the-power-of-personalization/?sh=290e19092a94).
8. *Social media customer service software*: Once upon a time, Facebook, Instagram, X (Twitter), and other social media platforms were simply digital homes to post pictures of food you had eaten and to argue with strangers about politics. Though those are still primary uses of the platforms, more people are trying to resolve their issues via social customer service. The common suggestion is to "meet your customers where they already are." Though that sounds like good advice, every team has limitations on time and resources, so it is important to create a social media customer service strategy that matches your team's budget and goals. For some, that might mean using your marketing team's existing social media management platform to handle customer inquiries, while for others, it might be the use of software integrations to bring social networks into your help desk. Beyond basic request management, social monitoring software can also be a great social media customer service tool. It helps you watch out for mentions of your company, competitors, and industry on social channels, giving you a heads up to issues so that they can be handled proactively.
9. *Service desk software*: Sometimes referred to as a ticketing system, service desk software is more IT-focused and is used to manage internal requests for technical assistance. It is not software that your customer service team would use directly, but it can be important for managing certain customer issues. For example, many teams use a ticketing system to manage bugs reported by customers. So even though your customer service team is not managing conversations directly in the tool, it is very common for them to have some interaction with it.

WHAT ARE SOME CUSTOMER SUPPORT SOFTWARE TRENDS?

Regardless of the type, software is always evolving. Sometimes the changes are due to shifts in customer or industry expectations. Other times, updates stem from technological advances that allow developers to offer features that were not previously possible. Some of the latest trends in CS software are identified here:

1. *Personalized conversations over ticket numbers*: While the focus of conversations over tickets has been a guiding principle for some customer service platforms since their start, the idea has become more mainstream in recent years. Customers are not interested in stiffly written scripts or one-size-fits-all email templates that never quite fit the issue at hand. Instead, customers want to have conversations with businesses where their concerns and needs are listened to and met in a timely manner.
2. This push for personalized support makes it even more important to choose a tool that gives your team access to customer information like past conversations and order history. Having customer data on hand ensures that they do not need to repeat themselves, always receive relevant information, and never feel like a number.
3. *Self-service continues to be a customer favorite*: A 2020 Zendesk report found that "69% of customers want to resolve as many issues as possible on their own and 63% of customers always or almost always start with a search on a company's online resources when they have an issue." (https://d1eipm3vz40hy0.cloudfront.net/pdf/cxtrends/cx-trends-2020-full-report.pdf). Customer service software that includes a knowledge base builder will allow you to create online resources that help customers find answers on their own while also relieving stress on your team. Another self-service feature found in support software is chatbots. Rule-based chatbot experiences can help customers get help more quickly. While not suited for complex issues, chatbots can often help with issues like providing tracking information and processing returns and exchanges.
4. *The usage of AI has proved to improve customer satisfaction and productivity*: With the recent updates to ChatGPT, most customer support platforms have started to offer AI features built into their products (https://www.helpscout.com/blog/chatgpt-customer-service/). Though their introduction may make your support team concerned about the future of their employment, the fact is that we're not quite there yet. Though the improvements to ChatGPT are quite impressive, their accuracy levels are still too low to be used without human supervision. Still, the technology offers many opportunities to help your team behind the scenes. Look for AI features that can perform helpful tasks like summarizing long discussions or performing sentiment analysis to aid in the quality assurance process. And yes, AI can also help your team come up with potential answers to customer issues – just make sure that a human reviews them for accuracy before hitting send!
5. *Understand the job you are trying to do*: Are you building a high-touch, hand-holding service experience for a select group or a mass-volume, fast-turnover retail service? Different software solutions suit different environments. Therefore:
   a. *Assess your resources*: If you are a team of one or two, you cannot cover every support channel all day. And if you have a limited budget, there is no point in looking at the more expensive systems.

b. *Refine the list of possible options*: Knowing what you must work with and what you want to get done, narrow in on the most likely categories of customer service software you will use. Perhaps you want a shared inbox, a knowledge base, and live chat?
c. *Understand the "must have" features*: Does this system absolutely have to integrate with an existing tool? Or is having a messaging system essential? That will help you narrow down your options.
d. *Create a shortlist*: Using those must-have features, reviews, recommendations, and other sources of insight, pick your top few options.
e. *Evaluate your favorites*: Now you can deep dive into your top few options, perhaps trying out their customer service and talking to existing customers.
f. *Trial time!* Using customer service software is the best way to know if it will work for you. All the feature checklists and marketing copy in the world won't replace the experience of using the software in your setup to deliver service.

Epilog

This book is essentially about how to achieve excellence in customer and supplier relationships. It is not about excellence in manufacturing or service, but achieving excellence in the entire organization for both customer and supplier. Organizations must realize and implement a win-win relationship for a lifetime relationship. This means that the relationship of excellence is measured in loyalty and respect for each other, and repeat business. In this relationship, quality is important; however, this quality must be at all levels of the organization. After all, quality is achieved when *customers return, but products do not*! This means, in the simplest definition, that quality is about *survival* in the future among competitors. So, what sets an organization apart?

We hope that we have demonstrated throughout this book that quality is not about standards, measurement, special characteristics, or special techniques. These, of course, are important, but most important is the *commitment to process excellence* and an overall desire *to do more* than just deliver *conformance to requirements* or products/services that are *fit for use*.

The start of differentiation begins with a flawless launching. To have a successful launching for any product and or service, both the organization and supplier must have focus and determination for a mutual engagement that delivers supplier components/systems on time with correct designs, proven performance, best-in-class quality, competitive cost, and required capacity. This means that a simple model must be followed consistently with the following five characteristics:

1. Planning
2. Implementation
3. Measurement
4. Improvement
5. Recognition and award

In practice, these five characteristics may be summarized as:

- *Early involvement*: Work with suppliers more *consistently*, earlier in the product development process.
- *Teams*: Utilize cross-functional teams to work together with suppliers at their facilities.
- Standardization: Adopt a single, global, disciplined process to minimize rework and redundancy.
- *Governance and reporting*: Use common, clear deliverables and a structured management review cadence.
- *Priority*: Have a system that prioritizes the needs for an intervention based on criticality, delivery, performance, capacity, or some other factor.

An excellent example of these five characteristics is the relationship of Ford Motor Company and two of its suppliers for the 2015 model of the F-150 truck. The first is the

TABLE E.1
Summary of Growth in Quality Perception

	Old Approach	New Approach
Quality services	• Service quality and performance • Customer satisfaction • Service quality applications	• Service management principles • Customer choices • Service failures and guarantees • e-service
Quality management	• Theory building • Cost of quality • TQM • Infrastructural element • Externalities • Global issues	• Usage of models • Context and contingencies • Global contingencies • Six sigma • Supply chain quality • Environmental concerns
Quality product	• Design based on attributes and performance • Product strategy • Process strategy	• Strategic product choices • New product introductions
Quality process	• Project management • Process models	• Improvement models based on data • Use statistics for optimization
Quality performance	• Quality practices • Quality awards	• TQM and performance • Accountability and measurement • Supply chain development • Usage of information technology • Focus on strategic benefits
Quality practice	• Theoretical models • Change management • Baldrige Practices • Learning contingencies	• Build on knowledge management (things learned – corporate knowledge) • Accumulating knowledge (historical TGW and TGR) • Baldrige Practices • SPC • Six sigma • ISO standards • Lean Practices

Alcoa Company, which provided aluminum for the truck body, and the second is the OSRAM Automotive lighting, which provided a complete LED forward lighting system.

The relationship of both customers and suppliers in this case demonstrates the growth in quality perception, which is summarized in Table E.1.

Both suppliers, Alcoa and OSRAM, were committed to satisfy the customer, and they both demonstrated this commitment through four key principles.

1. *Quality*: We all know how much consumers rely on the products they buy to make their lives easier and more comfortable, so both suppliers pushed themselves to make cars lighter, last longer (aluminum), and perform better (lighting system).

2. *Safety*: Everyone knows that as long as human beings design "whatever," there is going to be a failure, for nothing is perfect and absolute. In fact, from a reliability perspective, we know this from the R(t) = 1 –F(t) [reliability at time t is equal to 1 minus the failure rate]. So, the focus was to produce products that assured the customer was indeed buying a product that met and or exceeded the requirements of safety and that the new innovations were part of the philosophy of continual improvement in both cost and safety.
3. *Support and satisfaction*: Both customers and suppliers worked together in educating each other about the "new" innovation or bad experience at any level, to optimize product performance and value to consumers. (The assumption here is that any non-conformance – small, or big – will cause the customer to complain).
4. *Innovation*: Although aluminum usage in the automotive industry is not new, it is the first truck that has annual sales greater than any other vehicle on the market to be made of anything other than steel. The old model worked so well that the F-150 has been awarded the #1 status in its category for many years. However, the management of Ford was not satisfied with that. They wanted to improve. So, they introduced the concept of aluminum and LED technology. Indeed, in the F-150 for the 2015 model year, they incorporated both these innovations. The aluminum did contribute to the reduction of weight, thereby increasing fuel efficiency (miles per gallon), and the LED technology did contribute to much greater safety for the high/low beam, light driver module, turn module, and park module.

THE FIVE CRITICAL ITEMS FOR THE CUSTOMER

It all starts with customer *needs* and *wants*. After all, if there is no customer, there is no need for a supplier. We emphasize the word *supplier* because unless the relationship between customer and supplier is based on a win-win philosophy neither one will prosper. This win-win philosophy is based on a long-term relationship and is totally opposite to the *vendor* concept, which is based on short-term relationship and strictly on price. [a.n. An excellent example of this relationship was the Frito-Lay relationship after World War II.] So, in order for the customer to receive what they need and want, the following five critical items must be understood and implemented for a flawless launching of their product or service.

1. *Develop a formal enterprise risk*: This means that it is the customer's responsibility to identify supply chain risk and plan accordingly.
2. *Evaluate the supplier base for vulnerability and risk exposure*: Based on the results of the development of a formal enterprise risk, it is the customer's responsibility to analyze the individual supplier for their ability to perform to the needed requirements. Special consideration should be given to historical performance, capability, and capacity.
3. *Evaluate logistical issues*: It is profoundly important for the customer to make sure that logistics will not interfere with the delivery of the product and or service. Issues of concern here may be: Availability of utilities; Qualified

employment pool; Proximity to the facility of the customer; Transportation issues (availability of rail, airports, and highway system), and so on.

4. *Develop a contingency plan*: Things happen, and sometimes the results are devasted. Therefore, it is the customer's responsibility to develop contingency plans in case of emergencies, *i.e.*, acts of God, power failures, employee strikes, sudden increases in demand for the product or service, and so on.
5. *Develop a symbiotic relationship in the name of CI and win-win philosophy*: Quite often, the supplier may need help in financing to support their business, and or develop their business with new innovations. The customer should have planned for such occasions and be prepared to help. [In the case of the F-150, Ford helped the suppliers in the development of both grants and research.] In addition to direct help, the customer may want to encourage the supplier to pursue innovation and implementation of newer technologies. This is what supplier development is all about.

THE TEN INDICATORS FOR A RESPONSIBLE SUPPLIER

No one will deny that suppliers are a key component to every customer's success. After all, they are an integral part of the customer's quality, cost, and ultimate consumer satisfaction. Therefore, due to the significance of their contribution, we have identified some key indicators for a supplier who has the potential of being an "excellent" partner in the customer-supplier relationship. The indicators are:

1. Does it meet the standards and specific requirements?
2. Will it make it difficult for us to create more and better-quality products/services?
3. Does it use accounting tricks to mask more realistic spending or revenue figures (scrap, rework, capacity, *etc.*)?
4. Does it increase the price of the product/service?
5. Does it increase the cost of doing business with this particular organization?
6. Does it increase the cost of doing business directly or due to heavy regulations/specifications or policies?
7. Does it secure safety and sustainability attitude in what they do?
8. Does it support continual improvement for their own improvement, or is it viewed as a requirement of the customer?
9. Is the entire organization on board with the concept of continual improvement?
10. Are the policies and procedures in tandem with the philosophy of the customer?

These ten items we believe will signal to a customer whether or not the supplier is committed to improvement for its own sake and whether or not they are willing to perform on the path of excellence.

An analysis of the ten items will confirm early on in the selection process, the attitude of the management and the future relationship. Specifically, a cursory overview of these items will lead to a deeper discussion of the following:

- *Financial numbers*: Is the supplier financially sound to undertake the responsibility of supplying products and services to the customer?

- *Commitment to the customer*: Is the supplier committed to CI for all that it does for the customer? Does the supplier have enough capacity to satisfy the customer's needs?
- *Measurement ability*: Does the supplier have the ability to confirm the performance needed by the customer? How are the key performance indicators (KPI) identified? Measured? Recorded? Are they relevant to the customer needs? Are they actionable?
- *Plans and priorities*: Does the supplier have plans for CI? How are priorities set? Are the plans and priorities reviewed on a regular basis for changes and or adjustments?
- *Management plans*: Does the supplier cascade the plans and priorities to the appropriate level for action? Does management review the Quality Operating System (QOS) monthly?
- *Tracking the progress*: How does the supplier track the progress? How is progress measured? Are measurement data used, or are most of the measurements of the attribute type? Is SPC used by the operators? Reviewed by engineers, supervisors, and appropriate action taken based on the data?

A final thought about the win-win relationship of the customer and supplier may be summarized in seven principles that, if followed, the relationship will be fruitful and ready to complete a flawless launching. These principles are:

1. *Make the products easy to find.* That is the sourcing advantage. The closer the suppliers are, the better the flow and delivery will be to the customer. [a. n. This was the secret ingredient of Frito-Lay's success. The innovation to promote the product at the point of payment.]
2. *Get good advice.* Research within and outside of your organization for the **best** alternatives available. Benchmarking and Quality Function Deployment (QFD) are appropriate.
3. *Pick the right suppliers.* A customer depends on suppliers for the quality of their products. Make sure that the suppliers are able and willing to perform up to your expectations. They are your partners in quality. Treat them as equals.
4. Underst*and supplier rules and regulations.* Suppliers are also in business to make money. Therefore, respect their rules, culture, and regulations which they must work under. If they do not make money in the relationship, they will not be your supplier for long.
5. *Communicate regularly and efficiently with your suppliers.* It is imperative that there must be a two-way communication between the two. The information should be on requirements and expectations, as well as feedback. The feedback should be objective and data-driven!
6. *Watch out for channel conflict.* To be sure, conflicts will arise. It is imperative that both have a system of conflict resolution that is amicable to both parties.
7. *Look for continual improvement.* The improvement should be on several fronts, *i.e.*, communications, expectations, specifications, capability, capacity, costs, safety, quality, delivery, and in any area that is of concern.

Appendix A: A Summary of Several CS Software Based on https://www.helpscout.com/playlists/customer-service-software/

Customer Service Software	Starting Price	Supported Channels	Key Features	Best for
Help Scout	Free trial available. Plans start at $20/user per month	• Email • Contact form • Live chat • Social media • Knowledge base	• Shared inbox • Workflow automations • AI tools • In-app messaging • Integrations • Reporting dashboards	Growing businesses
HelpDocs	Free trial available. Plans start at $55 per month	• Knowledge base	• WYSIWYG editor with markdown and HTML support • Team accounts w/user roles • Article versioning • Integrations • Reporting dashboards	Stand-alone knowledge bases
Olark	Free trial available. Plans start at $29/seat per month (with one-year contract)	• Live chat	• Round-robin routing • Searchable transcripts • Chatbots • In-app messaging • Reporting dashboards	Stand-alone live chat
Aircall	Free trial available. Plans start at $30/license per month (minimum three users)	• Voice • SMS (text) messaging	• Shared call inbox • Interactive voice response (IVR) • Call transcripts • Integrations • Reporting dashboards	Call centers

(Continued)

Customer Service Software	Starting Price	Supported Channels	Key Features	Best for
Zendesk	Free plan and trial available. Plans start at $19/agent per month *Some channels and features only available via Zendesk Suite plans	• Email • Contact form • Voice • SMS (text) messaging • Social media • Live chat/messaging • Knowledge base • Community forums	• Omnichannel support • Workflows and automations • Chatbots • Integrations • Reporting dashboards	Larger teams
Freshdesk	Free plan and trial available. Paid plans start at $15/agent per month *Some channels and features only available via the omnichannel plan	• Email • Contact form • Voice • SMS (text) messaging • Social media • Live chat • Knowledge base • Community forums	• Shared inbox • Round-robin routing • Chatbots • Integrations • Reporting dashboards	Enterprise teams
Text-Em-All	Free plan available. Paid plans start at $10 per month	• SMS (text) messaging • Voice	• Templates • Surveys • Live transfers • Integrations • Reporting dashboards	Automated voice communication
Buffer	Free individual plan and trial available. Paid plans start at $5 per month	• Social media	• Post scheduling • Approval workflows • Hashtag manager • Custom link shortener • Reporting dashboards	Social media management
Helpshift	Free trial available. Plans start at $150 per month (includes unlimited agent seats)	• Email • Customer portal/ contact form • Live chat • Knowledge base	• Shared inbox • Chatbots • Web and in-app messaging • Integrations • Reporting dashboards	In-app support

(Continued)

Customer Service Software	Starting Price	Supported Channels	Key Features	Best for
HubSpot Service Hub	Free plan available. Paid plans start at $18 per month (includes two user seats)	• Email • Customer portal/ contact form • Live chat • Social media • Voice • Knowledge base	• Shared inbox • Chatbots • Meeting scheduling • Integrations • Reporting dashboards	Current Hubspot customers
Salesforce Service Cloud	Free trial available. Plans start at $25/user per month *Some channels and features only available on higher-tiered plans or as add-ons	• Email • Customer portal/ contact form • Live chat • Social media • SMS (text) messaging • Knowledge base • Community forums	• Shared inbox • Workflow automations • Chatbots • Integrations • Reporting dashboards	Slack support
LiveAgent	Free plan and trial available. Paid plans start at $9/agent per month	• Email • Customer portal/ contact form • Live chat • Social media • Voice • Video calling • Knowledge base • Community forums	• Shared inbox • Rule-based automations • Gamification • Integrations • Reporting dashboards	Teams interested in gamification
ServiceNow	Contact for quote	• Email • Customer service portal/contact form • Live chat • Social media Messaging • Voice • Knowledge base	• Shared inbox • Chatbots • Field service and asset management • Integrations • Reporting dashboards	Enterprise company IT teams
Zoho Desk	Free plan and trial available. Paid plans start at $7/user per month	• Email • Contact form • Live chat • Social media • Messaging • Voice • Knowledge base	• Shared inbox • Workflow automations • Chatbots • Integrations • Reporting dashboards	Sales-focused teams

(Continued)

Customer Service Software	Starting Price	Supported Channels	Key Features	Best for
Jira Service Management	Free plan available. Paid plans start at $22.05/ agent per month	• Email • Self-service portal/ contact form • Live chat • Knowledge base	• Shared inbox • Incident alerts • Asset and service management • Integrations • Reporting dashboards	IT teams
Sprout Social	Free trial available. Plans start at $249 per month. Additional users can be added for $199/user per month	Social media	• Post scheduling • Approval workflows • Shared inbox • Social listening tools • Reporting dashboard	Enterprise social media management

Appendix B: Various Specific Tools That May Be Used in the Process of CS and CI

Activity Network: The Activity Network is generally used to schedule dependent activities within a plan. It can be used for describing and understanding the activities within a standard work process. The resulting diagram is useful for communicating the plan and risks to other people.

When to use it:

- Use it when planning any project or activity that is composed of a set of interdependent actions.
- Use it to calculate the earliest date the project can be completed, and to find ways of changing this.
- Use it to identify and address risks to completing a project on time.

Affinity Diagram: The Affinity Diagram provides a visual method of structuring a large number of discrete pieces of information.

When to use it:

- Use it to bring order to fragmented and uncertain information, and where there is no clear structure.
- Use it when information is subjective and emotive, to gain consensus whilst avoiding verbal argument.
- Use it when current opinions, typically about an existing system, obscure potential new solutions.
- Use it, rather than a Relations Diagram, when the situation calls more for creative organization than for logical organization.

Bar Chart: Bar Charts are used to show the differences between related groups of measurements.

When to use it:

- Use it when a set of measurements can be split into discrete and comparable groups, to show the relative change between these groups.
- Use it when there are multiple sets of measurement groups, to show the relationship and change within and between groups.
- Use it, rather than a Line Graph, to display discrete quantities rather than continuing change.
- Use it, rather than a Pareto Chart, when a consistent ordering of bars is wanted. This can ease recognition and comparison of current and previous charts.

Brainstorming: Brainstorming is used to creatively generate new ideas. The creative synergy of a Brainstorming session is also useful in helping a team bind together.

When to use it:

- Use it when new ideas are required, to generate a large list of possibilities.
- Use it when a solution to a problem cannot be logically deduced.
- Use it when information about a problem is confused and spread across several people, to gather the information in one place.

Cause and Effect: The Cause-Effect Diagram is used to identify and structure the causes of a given effect. Use it in preference to a Relations Diagram where there is one problem and causes are mostly hierarchical (this will be in most cases).

When to use it:

- Use it when investigating a problem, to identify and select key problem causes to investigate or address.
- Use it when the primary symptom (or effect) of a problem is known, but possible causes are not all clear.
- Use it when working in a group, to gain a common understanding of problem causes and their relationship.
- Use it to find other causal relationships, such as potential risks or causes of desired effects.

Check Sheet: The Check Sheet is used to manually collect data in a reliable, organized way.

When to use it:

- Use it when data is to be recorded manually, to ensure the data is accurately recorded and is easy to use later, either for direct interpretation or for transcription, for example, into a computer.
- Use it when the recording involves counting, classifying, checking, or locating.
- Use it when it is useful to check each measurement as it is recorded, for example, that it is within normal bounds.
- Use it when it is useful to see the distribution of measures as they are built up.

Control Chart: The Control Chart is used to identify dynamic and special causes of variation in a repeating process. It is only practical to use it when regular measurements of a process can be made.

When to use it:

- Use when investigating a process, to determine whether it is in a state of statistical control and thus whether actions are required to bring the process under control.
- Use it to differentiate between special and common causes of variation, identifying the special causes that need to be addressed first.

- Use it to detect statistically significant trends in measurements; for example, to identify when and by how much a change to the process has improved it.
- Use it as an ongoing "health" measure of a process, to help spot problems before they become significant.

Decision Tree: The Decision Tree is used to select from a number of possible courses of action.

When to use it:

- Use it when making important or complex decisions, to identify the course of action that will give the best value according to a selected set of rules
- Use it when decision-making, to identify the effects of risks.
- Use it when making plans to identify the effects of actions and possible alternative courses of action.
- Use it when there are chains of decisions, to determine the best decision at each point.
- Use it only when data is available, or can reasonably be determined, on costs and probabilities of different outcomes.

DOE: "Design of Experiments" is used to understand the effects of different factors in a given situation.

When to use it:

- Use it when investigating a situation where there are many variable factors, one or more of which may be causing problems.
- Use it when variable factors may be interacting to cause problems.
- Use it when testing a solution, to ensure that there are no unexpected side effects.
- Use it when there is no time or money to try every combination of variables.

FMEA: FMEA is used to identify and prioritize how items fail, and the effects of failure.

When to use it:

- Use it when designing products or processes, to identify and avoid failure-prone designs.
- Use it when investigating why existing systems have failed, to help identify possible causes and remedies.
- Use it when investigating possible solutions, to help select one with an acceptable risk for the known benefit of implementing it.
- Use it when planning actions, in order to identify risks in the plan and hence identify countermeasures.

Fault Tree: Fault Tree Analysis is used to show combinations of failures that can cause overall system failure.

When to use it:

- Use it when the effect of a failure is known, to find how this might be caused by combinations of other failures.

- Use it when designing a solution, to identify ways it may fail and consequently find ways of making the solution more robust.
- Use to identify risks in a system, and consequently identify risk reduction measures.
- Use it to find failures which can cause the failure of all parts of a "fault-tolerant" system.

Flowchart: The Flowchart is used to show the sequential steps within a process.

When to use it:

- Use it when analyzing or defining a process, to detail the actions and decisions within it.
- Use it when looking for potential problem points in a process.
- Use it when investigating the performance of a process, to help identify where and how it is best measured.
- Use it as a communication or training aid, to explain or agree on the details of the process.

Flow Process Chart: The Flow Process Chart is used to record and illustrate the sequence of actions within a process.

When to use it:

- Use it when observing a physical process, to record actions as they happen and thus get an accurate description of the process.
- Use it when analyzing the steps in a process, to help identify and eliminate waste.
- Use it, rather than a Flowchart, when the process is mostly sequential, containing few decisions.

Force Field Analysis: The Force Field Diagram is used to weigh up the points for and against a potential action.

When to use it:

- Use it when decision-making is hindered by a number of significant points for and against a decision.
- Use it when there is a lot of argument and indecision over a point, to clarify and agree on the balance of disagreement.
- Use it to help identify risks to a planned action and to develop a strategy for counteracting them.
- Use it to help identify the key causes of successful or unsuccessful actions.

***Gant Chart*:** Gantt Charts are used to show the actual time to spend on tasks.

When to use it:

- Use it when doing any form of planning, to show the actual calendar time spent in each task.
- Use it when scheduling work for individuals, to control the balance between time spent during normal work hours and overtime work.

- Use it when planning work for several people, to ensure that those people who must work together are available at the same time.
- Use it for tracking progress of work against the scheduled activities.
- Use it when describing a regular process, to show who does what, and when.
- Use it to communicate the plan to other people.

Histogram: The Histogram is used to show the frequency distribution of a set of measurements.

When to use it:

- Use it to investigate the distribution of a set of measurements.
- Use it when it is suspected that there are multiple factors affecting a process, to see if this shows up in the distribution.
- Use it to help define reasonable specification limits for a process by investigating the actual distribution.
- Use it when you want to see the actual shape of the distribution, as opposed to calculating single figures like the mean or standard deviation.

IDEFO (ICOM DEFinitions): Integration definition for Process Modeling (IDEF0) is used to make a detailed and clear description of a process or system.

When to use it:

- Use it when formally describing a process, to ensure a detailed, clear, and accurate result.
- Use it when the process is complex, and other methods would result in more complex diagrams.
- Use it when mapping a wide variety of processes, as a consistent and scalable Process Description Language (or PDL).
- Use it when there is time available to work on understanding and producing a complete and correct description of the process.

Kano Model: Kano analysis is a quality measurement tool used to prioritize customer requirements based on their impact on customer satisfaction.

When to use:

- Kano analysis is a tool that can be used to classify and prioritize customer needs. This is useful because customer needs are not all of the same kind, not all have the same importance, and are different for different populations. The results can be used to prioritize your effort in satisfying different customers.

Kappa Statistic: The Kappa is the ratio of the proportion of times the appraisers (see Gage R&R) did agree to the proportion of times the appraisers could agree. If you have a known standard for each rating, you can assess the correctness of all appraisers' ratings compared to the known standard. If Kappa = 1, then there is perfect agreement. If Kappa = 0, then there is no agreement. The higher the value of Kappa, the stronger the agreement. Negative values occur when agreement is weaker than expected by chance, but this rarely happens. Depending on the application, Kappa less than 0.7 indicates that your measurement system needs improvement. Kappa values greater than 0.9 are considered excellent.

When to use it:

- Use it when there is a need for comparison.

Line Graph: The Line Graph is used to show patterns of change in a sequence of measurements.

When to use it:

- Use it when an item is repeatedly measured, to show changes across time.
- Use it when measuring several different items that can be shown on the same scale, to show how they change relative to one another.
- Use it when measuring progress toward a goal, to show the relative improvement.
- Use it, rather than a Bar Chart, to show continuous change, rather than discrete measurements. It is also better when there are many measurements.
- Use it, rather than a Control Chart, when not measuring the degree of control of a process.

Matrix Data Analysis Chart: The Matrix Data Analysis Chart (MDAC) is used to identify clusters of related items within a larger group.

When to use it:

- Use it when investigating factors that affect a number of different items, to determine common relationships.
- Use it to determine whether or not logically similar items also have similar factor effects.
- Use it to find groups of logically different items that have similar factor effects.

Matrix Diagram: The Matrix Diagram is used to identify the relationship between pairs of lists.

When to use it:

- Use it when comparing two lists to understand the many-to-many relationship between them (it is not useful if there is a simple one-to-one relationship).
- Use it to determine the strength of the relationship between either single pairs of items or a single item and another complete list.
- Use it when the second list is generated as a result of the first list, to determine the success of that generation process. For example, customer requirements versus design specifications.

Nominal Group Technique: The Nominal Group Technique (NGT) is used to collect and prioritize the thoughts of a group on a given topic.

When to use it:

- Use it when a problem is well understood, but knowledge about it is dispersed amongst several people.

- Use it when a rapid consensus is required from a team, rather than a more detailed consideration.
- Use it when the team is stuck on an issue, for example, when they disagree about something.
- Use it when the group prefers a structured style of working together.
- Use it, rather than Brainstorming, when a limited list of considered opinions is preferred to a long list of wild ideas, or when the group is not sufficiently comfortable together to be open and creative.

Pareto Chart: The Pareto Chart is used to show the relative importance of a set of measurements.

When to use it:

- Use it when selecting the most important things on which to focus, thus differentiating between the 'vital few' and the 'trivial many'.
- Use it after improving a process, to show the relative change in a measured item.
- Use it when sorting a set of measurements, to visually emphasize their relative sizes.
- Use it, rather than a Bar Chart or Pie Chart, to show the relative priority of a set of numeric measurements.

Prioritization Matrix: The Prioritization Matrix is used to sort a list of items into an order of importance.

When to use it:

- Use it to prioritize complex or unclear issues, where there are multiple criteria for deciding importance.
- Use it when there is data available to help score criteria and issues.
- Use it to help select items to be actioned from a larger list of possible items.
- When used with a group, it will help to gain agreement on priorities and key issues.
- Use it, rather than simple Voting, when the extra effort that is required to find a more confident selection is considered to be worthwhile.

Process Capability: Process Capability calculations indicate the ability of a process to meet specification limits.

When to use it:

- Use it when setting up a process, to ensure it can meet its specification limits.
- Use it when setting specification limits, to ensure they are neither too wide nor too narrow.
- Use it when investigating a process that is not meeting its specification limits.
- Use it only when the process is stable and has a normal distribution.

Process Decision Program Chart (PDPC): The Process Decision Program Chart (commonly just referred to as PDPC) is used to identify potential problems and countermeasures in a plan.

When to use it:

- Use it when making plans, to help identify potential risks to their successful completion.
- When risks are identified, use it to help identify and select from a set of possible countermeasures.
- Also use it to help plan for ways of avoiding and eliminating identified risks.
- It is of best value when risks are non-obvious, such as in unfamiliar situations or in complex plans, and when the consequences of failure are serious.

Relationship Diagram: The Relations Diagram is used to clarify and understand complex relationships.

When to use it:

- Use it when analyzing complex situations, where there are multiple interrelated issues.
- Use it where the current problem is perceived as being a symptom of a more important underlying problem.
- It is also useful in building consensus within groups.
- It is commonly used to map cause-and-effect relationships, but also can be used to map any other type of relationship.
- Use it, rather than an Affinity Diagram, when there are logical, rather than subjective, relationships.
- Use it, rather than a cause-and-effect Diagram, when causes are non-hierarchic or when there are complex issues.

Scatter Diagram: The Scatter Diagram is used to show the type and degree of any causal relationship between two factors.

When to use it:

- Use it when it is suspected that the variation of two items is connected in some way, to show any actual correlation between the two.
- Use it when it is suspected that one item may be causing another, to build evidence for the connection between the two.
- Use it only when both items being measured can be measured together, in pairs.

Simulation in Lean: Simulation is a practical tool that changes complex parameters into mathematical, logical models that are possible to analyze under different scenarios. Models can be classified from various aspects. They can be categorized as being mathematical or physical, static (Monte Carlo) or dynamic (shows a system that always changes), deterministic (includes no random variables) or stochastic (contains at least one random variable), and also discrete or continuous models. But the simulation, besides its benefits also has some disadvantages.

When to use:

- Assessing hospital operations in different wards.
- Estimating the number of needed doctors and nurses in the emergency ward.
- Reducing the length of time that patient has to stay in the emergency ward.
- Predicting the required number of beds according to the patient interval.

Advantages:

1. The most important benefit of simulation is to get results at less cost.
2. Assumptions about specific phenomena can be checked for feasibility.
3. The simulation answers "what if" questions.
4. Analysis of bottlenecks and finding the roots is easier.
5. The system can be tested without allocating any resources.

Disadvantages:

1. Working with simulation modeling requires experience and specific training.
2. Understanding the simulation results is difficult because they are based on random inputs.

String Diagram: The String Diagram is used to investigate the physical movement in a process.

When to use it:

- Use it when analyzing a manual or physical process that involves significant physical movement, in order to make movements easier and quicker. Movements may be of people, materials, or machines.
- Use it when designing the layout of a work area, to identify the optimum positioning of machines and furniture.

Surveys: Surveys are used to gather information from people.

When to use it:

- Use it when information that is required is held by an identifiable and dispersed group of people.
- Use it to help decision-making by turning disparate qualitative data into useful quantitative information.
- Use it only when the time and effort are available to complete the survey.
- Use it, rather than Brainstorming or Nominal Group Technique, to gather real data about what a diverse group of people think (rather than an opinion of what they think).

Tables: Tables are used to organize and relate multiple pieces of information.

When to use it:

- Use it when gathering information, to help prompt for a complete set of data.

- Use it when information is disorganized, to help collate and understand it.
- Use it for summarizing information to make it easier to present and understand.
- Use specific table tools as frameworks for particular tasks, either to organize existing information or to prompt for specific categories of information.

Tree Diagrams: The Tree Diagram is used to break down a topic into successive levels of detail.

When to use it:

- Use it when planning, to break down a task into manageable and assignable units.
- Use it when investigating a problem, to discover the detailed component parts of any complex topic.
- Use it only when the problem can be broken down in a hierarchical manner.
- Use it, rather than a Relations Diagram, to break down a problem when the problem is hierarchical in nature.

Value Analysis: Value Analysis is used to determine and improve the value of a product or process.

When to use it:

- Use it when analyzing a product or process, to determine the real value of each component.
- Use it when looking for cost savings, to determine components that may be optimized.
- Use it only when the item to be analyzed can be broken down into sub-components and realistic costs and values allocated to these.

Voting: Voting is used to prioritize and select from a list of items.

When to use it:

- Use it as a quick tool when a group must select one or more items from a list, for example, as generated by a Brainstorming session.
- Use it when it is important that the group accept the result as fair.
- Use it when the knowledge to enable selection is spread within the group.
- Use it when the opinion of all group members is equally valued.
- It can also be used to "test the water," to determine opinions without committing to a final selection.
- Use it, rather than a Prioritization Matrix, when the added accuracy of the Prioritization Matrix is not worth the extra effort.

Appendix C: Lean Deployment Guideline Sheet

Do You See This	Do You See This?	Do You See This	Do You See This
Visual Factory		**Pull Systems**	
Non-Lean	**Lean**	**Non-Lean**	**Lean**
No notification information-(Andon) Boards No/little evidence of Lean metrics used Variations in work performance Standard procedures not being followed Dirty, cluttered, messy work areas Messy bathrooms Materials piled everywhere Unmarked gauges and tools No min/max levels at lineside Empty shadow boards, tools missing, disorganized dirty floors, overflowing drawers, cabinets, and desks	Used notification/information (Andon) Boards Demonstrate use of key Lean metrics Standardized Work forms prominently displayed/workers knowledgeable Workers do the job the same way every time Sparkling, clean equipment, tools, work areas and bathrooms Designated area for empty/full containers Marked gauges and key equipment Min/Max levels clearly marked Shadow boards for tools that are used and maintained Shiny clean floor-painted	High level of inventory on lines and in plant-receiving and shipping Materials waiting and stacked at line side Delivery timing of materials unregulated Centralized schedules w/out Takt times Products "pushed" to the next area	Low level of inventory on lines and in plant Real Kanban system in use Material delivered frequently to lineside Use of small supermarket areas Localized/Customer demand with Takt times Use of Load Levelling
Evidence of Teams		**Returnable Container/Dunnage**	
No production data visible (or outdated) Lack of conference room for teams to use	Available team meetings areas Team data prominently displayed - Quality data - Production real time data - Problem solving evidence	Cardboard containers Wooden pallets Large number of parts/containers Equipment stability Little visible evidence of preventive maintenance process	Re-usable containers (plastic, metal) recycled between organization and suppliers Small number of parts/containers Maintenance schedule posted and carried

(Continued)

Do You See This	Do You See This?	Do You See This	Do You See This
Visual Factory		**Pull Systems**	
Non-Lean	**Lean**	**Non-Lean**	**Lean**
Suggestion campaign old, suggestion box not active, little implementation of suggestions accepted	Training depth chart	Frequent breakdowns and no tracking	out – employees involved
No problem-solving involvement at operator level	Teams implemented many of their own suggestions	No Overall Equipment Effectiveness (OEE) data	Breakdowns fixed quickly and prevented
8D done in quality department	Proof of problem-solving activities in many places in plant (Fishbone diagrams filled out and displayed, PDCA cycle being followed, 5 Why's)	No study/analysis into major losses	OEE at 85% or better (posted and tracked by teams)
Team not actively involved in Standardized Work activity or workflow	Aggressive team Standardized Work input		
	Six Sigma is used to reduce variation		
Changeover		**Layout and Good Material Flow**	
Usually done by one person	Changeover done by team	Insufficient, poorly spaced docks	Point of use shipping and receiving docks
No tracking of changeover time	Clear internal/external task identification	Cluttered staging areas	Visually clear and simple staging areas
Most work is done when machine is down	Changeover chart tracks times, improvements, goals, ideas	Warehouse/Storage poorly placed for continuous flow	Well placed small market area
Lack of standardized methods - no specific work task/ procedures identified	Use of changeover chart, visual organization used, standard methods	Designated scrap areas that are full	Little scrap/use of mistake proofing
		Long conveyors, full of WIP	Short conveyors, min/max levels marked
		Cell Design	
		Inefficient assembly configurations	U-shaped cells/no wasted motion
		Assembly long way from point of use	Cells adjacent to point of use
		Storage of completed materials (more than 1 day supply)	Sub-assembly and raw materials storage of 4 hours to support production

(Continued)

Do You See This	Do You See This?	Do You See This	Do You See This
Visual Factory		Pull Systems	
		Lean Process Flow	
		Operation running much faster than customer requirements Large buffers/ production push product Large off-line repair bays	Operation tied to customer requirements Small buffers used sparingly to de-couple major processes Production pulled from prior operation

FIRST TIME THROUGH (FTT)

Definition: First Time Through is the percentage of units that complete a process and meet quality guidelines the first time without being scraped, rerun, retested, diverted for off-line repair, or returned. It is a measure of the quality of the manufacturing process.

$$\frac{\text{Units entering process} - (\text{scrap} + \text{reruns} + \text{retests} + \text{repaired off} - \text{line} + \text{returns})}{\text{Units entering process}}$$

Stretch Objective: One hundred percent (100%) FTT capability = Zero defects made or passed on.

Why:

Improvements in First Time Through will result in:

- Increased capacity.
- Improved product quality to internal and external customers.
- Reduced need for excess production inventory, resulting in improved Dock to Dock time.
- Improved ability to maintain sequence throughout the process, resulting in improved Build to Schedule performance.
- Increased quality input to constraint operations, resulting in improved Overall Equipment Effectiveness.
- Elimination of wastes due to scrap, repair, and excess inventory, resulting in improved Total Cost.

Where:

FIT data should be collected at the end of the processes.

When:

First Time Through data should be collected and used, at minimum, by shift. FTT data must drive business practices behavior.

First Time Through Calculation

Units entering process – (scrap + returns + repaired off-line + retests) Units entering process

Units entering	1000
Scrap	10
Returns	15
Retest	5
Repair off line	0
FTT%	1000–(10+15+5+0)/1000
FTT%	970/1000
FTT%	97.0%
FTT Multiple Operations Calculation Examples	
FTT – Operation 1	92.87%
FTT – Operation 2	87.65%
FTT –Operation 3	82.34%
Final Inspection – Operation 4	82.34%
Total FTT%	(.9287x.8765x.6598x.8234)
Total FTT%	44.22%

DOCK-TO-DOCK (DTD)

Definition: Dock-to-Dock is the elapsed time between the unloading of raw materials and the release of finished goods for shipment. DTD measures how fast raw materials are converted to finished goods and shipped. It is a measure (in hours) of the speed of material through the plant, not the speed of the processes.

Why:

Plants must eliminate waste and move towards customer "Pull" systems.

By building large buffers, you produce more parts (increase the speed of your process) but slow down the speed of materials.

Decreasing inventories leads to less material handling and storage, which results in fewer opportunities to damage parts.

Where:

Raw Material – Purchased Part (control part) – Begin counting when it is received in the plant, even on consignment, because it is already in its product form.

Raw material that is changed to its product form.

Work-in-Process (WIP), including material in repair/rework areas, and all buffers.

Finished Goods End Item Products – All finished goods containing the control part are included in the DTD calculation until they are physically shipped from the dock.

When:

Data should be collected at the same time that best represents the operating pattern of a product line.

TOTAL DOCK DO DOCK CALCULATION

DTD (Single Operation) = Total Units of Control Part / End of Line Rate

DTD (Multiple Operation) = DTD (Operation 1) + DTD (Operation 2) + DTD (Operation 3)... / End of Line Rate

The End of Line Rate is calculated off the last operation in the service line for a product:

Total Unit Produced	730 (from last operation)	
Production Hours	12 hours (10 scheduled production hrs = 2 unplanned overtime hrs)	
End of Line Rate	730/12 = 60.83	
Units in Area	Raw Material + Work-in-Process (by Operation) + Finished Goods	
Raw Material	300	
Operation 1	181	
Operation 2	3	
Operation 3	3	
Finished Goods	200	
DTD (Raw Material)	300/60.83	4.93 hrs
DTD (Operation 1)	181/60.83	2.98 hrs
DTD (Operation 2)	3/60.83	0.05 hrs
DTD (Operation 3)	3/60.83	0.05 hrs
DTD (Finished Goods)	200/60.83	3.29 hrs

Total DTD = 4.93 hr + 2.98 hr + .05 hr + .05 hr + 3.29 hr

Total Dock to Dock time = 11.3 Hours

OVERALL EQUIPMENT EFFECTIVENESS (OEE)

Definition: OEE is a combination of three different calculations. It is a measure of the availability, performance efficiency, and quality rate of a given piece of equipment. OEE data will help identify several wastes caused by machine/process inefficiencies.

Stretch Objective: Most common recommended standard, 85% OEE. The best OEE is 100%; however, it is not a realistic goal.

Why:

Higher throughput (increases capacity) decreasing Total Dock to Dock times.

More stable processes improve production predictability, therefore improving Build to Schedule.

Higher throughput and lower rework/scrap lead to improved First Time Through and Total Cost.

Where:

OEE should be measured on machines in your constraint operations.

When:

Data should be collected by shift and by product or job if applicable. OEE data should be reviewed daily and trended over time to identify improvement opportunities and verify the effectiveness of process changes.

Special consideration: OEE data must not be used to compare plant-to-plant performance. Too many variables exist in physical environments, facility layout, and machine/process uniqueness to compare on a valid basis.

OEE	Availability x Performance Efficiency x Quality
Availability	$\frac{\textit{Operating Time}}{\textit{Net Available Time}}$
Performance Efficiency	$\frac{\textit{Idea Cycle Time} - \textit{Total Parts Run}}{\textit{Operating Time}}$
Quality	$\frac{\textit{Total Parts Run} - \textit{Total Defects}}{\textit{Total Parts Run}}$

OEE CALCULATION EXAMPLE

Availability	
Total schedule time (including overtime)	720 min
Required downtime (*e.g.*, 30 min. lunch; 30 min breaks, etc.)	30 + 30 = 60 min
Net available time	720 – 60 = 660 min
All other downtime (*e.g.*, machine downtime and die changes)	45 + 45 = 90 min
Operating Time	660 – 90 = 570 min
Availability	$\frac{570}{660} = 86.3\%$
Performance Efficiency	
Ideal cycle time	0.33 min/part
Total parts run	1440
Operating time	570 min

(Continued)

Performance	(.33 x 1440) = 83.4%
Efficiency	570
Quality Rate	
Total parts run	1440
Rework	50
Scrap	40
Quality Rate	$\frac{1440-(50+40)}{1440}=93.8\%$
Overall Equipment Effectiveness	
Availability	86.3%
Performance efficiency	83.4%
Quality rate	93.8%
OEE	.863 x .834 x .938
OEE	67.5%

QUICK CHANGEOVER

Definition: A method for people to analyze and significantly reduce setup and changeover time.

Changeover Time: The time between the last good piece off one production run and the first good piece off the next production run after the changeover.

Internal: Activities performed while the machine is shut down.

External: Activities performed while the machine is safely running.

	Changeover Time					
1	Run A	external	Internal	External	Run B	
		Start: Document current changeover elements				
2	Run A	Internal		Run B		50% reduction
		Separate Internal/External Activities				
3	Run A	Internal	Run B			75% reduction
		Shift Internal Activity to External				
4	Run A	Internal	Run B			90% reduction
		Streamline Internal and or External Activities				

Quick Changeover Process:

- Document Current Changeover Process.
- Document each element (Consider Photos/Videos).
- Time the process and the elements.

Establish baseline

- Separate Internal and External Activities.
- Internal activities must be performed while the machine is down.
- External activities can be performed while the machine is running.
- Shift Internal Activities to External.

Analyze each element:

- Ask five (5) whys.
- Resolve if the element can be made external.
- Locate Parallel Activities.

Tasks that are independent of each other can be performed simultaneously

Changeover time can be reduced significantly by doing jobs in parallel (simultaneously)

Streamline Internal/External Activities:

- Adjustment elimination
- Fixtures and jigs
- Functional standards
- Quick die change hardware

Implement the Revised Plan:

- Examine feasibility of all ideas.
- Validate the Method/Verify the Results.
- Develop draft procedures.
- Test procedures.
- Verify results with data.
- Document the New Method on a Standard Worksheet.
- One Standard Worksheet for External.
- One for each set of parallel internal activities.
- Notify Other Areas.

VISUAL FACTORY

Definition: A facility in which anyone can know in 5 minutes or less the who, what, where, when, how, and why of any work area, without talking to anyone, opening a book, or turning on a computer.

VISUAL FACTORY LEVELS

Visual display

- *Level 1*: Share information. Sharing all information about activities lets everyone see how closely performance conforms to expectations.

- *Level 2:* Share standards at the site. Sharing information about standard specifications and methods lets everyone identify non-conformance as it occurs and helps to correct it.

Visual controls

- *Level 3*: Build standards into the workplace. Make the work environment (space limits, standard containers, equipment barriers) communicate established standards.
- *Level 4*: Warn about abnormalities. Incorporate lights, alarms, bells, buzzers, and other devices that alert workers when an abnormality is detected, such as a defect generated or a shortage of parts.
- *Level 5*: Stop abnormalities. Prevent defects from moving on after detection by stopping the process or rejecting the part. Devices are installed at the source of abnormalities.
- *Level 6*: Prevent abnormalities. Level 6 means 100% prevention of abnormalities. Once the exact cause of a defect is known, an error proofing device is installed to prevent the defect from occurring at all.

FIVE STEPS TO WORKPLACE ORGANIZATION

- Step 1: Sort (Organization)
  - distinguish between what is needed and what is not needed.
- Step 2: Stabilize (Orderliness)
  - "A place for everything and everything in its place."
  - Determine the best location for all necessary items.
- Step 3: Shine (Cleanliness)
  - Eliminating dirt, dust, fluids, and other debris to make the work area clean.
  - Adopting cleaning as a form of inspection.
- Step 4: Standardize (Adherence)
  - Maintain and monitor the first three "S's."
  - Check/Standardize/Maintain/Monitor/Improve.
- Step 5: Sustain (Self-discipline)
  - Correct procedures have become habit.
  - The workplace is well ordered according to agreed-upon procedures.
  - Managers, operators, and engineers are deeply committed to 5S.

ERROR PROOFING

Definition: Error Proofing is a process improvement to prevent a specific defect from occurring.

Three Sides of Error Proofing:

- Physical – Install hardware.
- Operational – Enforce procedure, sequence, and/or execution.
- Philosophical – Empowerment of workforce.

Approaches to Error Proofing

- Prevention – Prevents errors from creating defects.
- Detection – Detects defects and immediately initiates corrective action to prevent multiple defects from forming.
- *Definition of a Defect*: A defect is the result of any deviation from product specifications that may lead to customer dissatisfaction. To classify as a defect:
  - The product has deviated from manufacturing or design specifications.
  - The product does not meet internal and/or external customer expectations.
- *Definition of an Error*: An error is any deviation from a specified manufacturing process. There can be an error without a defect. There cannot be a defect without an error first. The result of an error is demonstrated in a defect.

Error-Proofing Approach

Error Proofing Teams: Error Proofing Teams are composed of operators, supervisors, engineers, skilled trades, and support personnel. The observations, abilities, and creativity of each individual are combined to:

- Make changes to the work environment.
- Increase production efficiency.
- Take ownership in the manufacturing process.
- Support continuous improvement in the workplace.

Error Proofing Method

- Identify the Problem.
- List Possible Errors.
- Determine the Most Likely Error.
- Propose Multiple Solutions.
- Evaluate Effectiveness, Cost, and Complexity of Solutions.
- Determine the Best Solution (Include Data Analysis).
- Develop Implementation Plan.
- Analyze Preliminary Benefits.
- Develop Plan for Long-Term Measure of Benefits.
- Congratulate the Team.

Benefits of Error Proofing

- Business costs: Cost of additional energy and equipment required to manage a poor quality production environment.
- Liability Costs: Cost of losing business due to poor product quality.
- Cultural Costs: Reduction of worker moral.

JUST IN TIME

Just-In-Time (JIT) establishes a system to supply work (data, information, *etc.*) or services (patient care) at precisely the right time, in the correct amount, and without error. JIT is the heart of a Lean system and is an overriding theme for Lean healthcare. JIT is attained through the understanding and application of continuous flow, the pull system, and Kanbans.

CONTINUOUS FLOW

Continuous Flow is characterized by the ability of a process to replenish a single unit of work (or service capacity) when the customer has pulled it. The concept of continuous flow is used to move work, patients, or provide a service between processes with minimal or no wait (queue) time. It is further used to ensure that the process is performing the work required, no sooner or no later than requested, as well as in the correct quantity, with no defects (nonconformances). The goal is not to do any work or service that is not requested by the downstream process (or customer). It is synonymous with JIT. By focusing on continuous flow, the team will be able to:

- Reduce or eliminate transport, delay, and motion waste.
- Decrease lead times.
- Reduce queue times.
- Allow staff to identify and fix problems earlier.
- Provide the needed flexibility in meeting demand changes.
- Improve patient and staff satisfaction levels.

True (100%) continuous flow in a manufacturing and service organization may not be achieved because of the staff and departmental interdependencies required in the various pathways of different processes. Therefore, the tools of In-process supermarkets, FIFO Lanes, and Kanbans can be used at certain times when cycle time differences between processes exist and flow needs to be improved.

IN-PROCESS SUPERMARKET

The In-process supermarket is a physical device between two processes that stores a certain quantity of work or service capacity that, when needed, is pulled by the downstream process. Its origination came from the grocery stores and how the clerks replenish food items. Supermarkets do not typically have large storage areas for many of their goods. Every item is received and transferred straight to the shelves (especially the perishable items), where it is made available to the customer. No costly storage. No large storage areas. No waiting. Minimum inventory. A good example of application in manufacturing is the tooling department. Tools need to be made on demand or have a very small inventory. Another example of in process supermarket is the hospital pharmacy for making IVs. Due to specific requirements and time life considerations, the IVs must be made practically on demand.

The In-process supermarket (*i.e.*, the grocery store shelf between the supplier of the goods and the consumer) exists due to the variations in customer demand. The customer's lead time is minimal (the time it takes to remove something from the shelf), therefore, it would not be feasible to have just one item there. In order to establish the In-process supermarket, an ordering pattern, establishing minimum and maximum levels, needs to be determined to create a balance between what the customer demand is and the frequency of delivery to the store (and shelf). In-process supermarkets can also be explained in terms of cycle time. A supermarket exists between two processes to accommodate the differences in the cycle times of those processes (*i.e.*, the time for the removal of a tool from the inventory to the time it takes to replenish that item). This is called the pull system. Only the amount that has been used by the customer for that day, week, or month (depending on the ordering pattern), or "pulled" from the shelf, is reordered. The maximum level is never exceeded. For In-process supermarkets to be successful, the following will need to be in place:

A quantitative understanding of the downstream requirements:

- Known cycle times for the downstream and upstream processes.
- Communication system between the upstream and downstream processes.
- Minimum and maximum number of work units or service capacity assigned to the supermarket.
- Standard work was created explaining the operation of the supermarket.
- Training of all process workers on the supermarket operation.
- Adequate documentation and visual controls communicating to the downstream process when there is a disruption in the flow.
- A signal or "kanban" triggers a "pull" from the downstream process to the upstream process to replenish what was removed

For example, supplies in many departments can run out on an hourly, daily, or weekly basis. To minimize this disruption and maintain optimal functions, In-process supermarkets for unit/floor supply areas may provide value. The supermarkets would limit extra supplies throughout the area and prevent the searching for needed supplies at critical times. The supplies at these locations would be located in an In-process supermarket, meeting the hourly and daily needs of the staff. Replenishment of the supplies would be through kanban cards when supplies reached a minimum ordering level.

It should be noted that information (*i.e.*, charts, labs, consults, *etc.*) can also be considered a "supply" or "supermarket item" and the supermarket concept can be applied as such. Many organizations (manufacturing, service, healthcare, financial, defense, *etc.*) have found the supermarket to be the best alternative to scheduling upstream processes that cannot flow continuously. As you improve flow, the need for supermarkets should decrease.

KANBAN

Kanban is a card or visual indicator that serves as a means of communicating to an upstream process precisely what is required at the specified time. In Japanese, Kanban means "card," "billboard," or "sign." Kanban refers to the inventory control card used in a pull system. It is used to regulate the flow or work in and out of supermarkets as a visual control to trigger action.

Kanban is a form of visual control (information that allows a process to be controlled). This information states when, who, what, and how many work units are needed for movement. A Kanban can be anything from an actual index card, a file folder, or some type of electronic signal. There needs to be a mailbox or some repository for the Kanban to be deposited in, as well as the signal system identifying it is there.

The most common use for the In-process supermarket is the ordering of supplies. Kanbans for supplies will ensure that the dollars allocated for supplies will be at the minimum required. The Kanban system is used to create a "pull" of material, in this case a supply item, from the downstream process to the upstream process. The formal application to Kanban is an eight-step approach:

Conduct the supply survey: The team will need to create a standard list of supplies from which to draw upon. Distribute the list to all staff in the area, requesting them to determine their usage level for a period of time (day/shift, week or month). Examples are: exam gloves, blue pads, general use syringes, and so on. A typical form will look like Table C.1.

Establish minimum/maximum levels: Once the list and special requests have been collected, gain a consensus on usage and establish minimum/maximum levels. Include the following:

- Type of standard supply (there are numerous glove sizes, so agreement must be made on a certain quantity for each size – small, medium, large, or one size fits all).
- Weekly or monthly usage for the items.
- Establish a minimum quantity to have on hand.
- Establish a maximum quantity to have on hand.

Create the Supply Order Form: The Supply Order Form will list all the supplies represented by information obtained from the Supply Survey Form – see Table C.2. The minimum quantity to have on hand is the quantity of items expected to be used during the time it takes to re-supply that item (plus some X factor or buffer). The X factor or buffer is to ensure no stock-out occurs at any time. The X factor is typically determined by the team and their experience (historical data) with the supplies. It may be one day's worth

TABLE C.1
A Typical Supply Survey Form

<table>
<tr><th colspan="3">Supply Survey Form</th></tr>
<tr><td colspan="3">Select from the list below the frequency of the items you will most likely use within a specific time (shift, week, month, and so on). If not on the list, please add the item at the bottom of the list. This info will be used in the Lean initiative to reduce or eliminate excess waste.</td></tr>
<tr><td>Item Description</td><td>Quantity Used/Time Period</td><td>Comments</td></tr>
<tr><td></td><td></td><td></td></tr>
<tr><td></td><td></td><td></td></tr>
<tr><td></td><td></td><td></td></tr>
</table>

TABLE C.2
A Typical Order Form

Supply Order Form				
Fax this form to the Central supply at extension XXXX by 16:00. Your supplies will be delivered by 06:30 the next day				
Number	Item description	Item #	Quantity ordered (Maximum – Minimum)	Unit price
1				
2				
3				

of a particular supply; it may be a week's worth, again depending on the experience of the team. The maximum quantity to have on hand will be the minimum quantity plus the number of items that would be used during the supply or replenishment time.

1. *Create the Kanban cards*: There should be one kanban card identified as the Supply Re-order Kanban for each supply item. It should be laminated and color-coded, differentiating more than one supply ordering location (*i.e.*, Production, Tool room, Central Supply, Housekeeping, *etc.*). Kanban cards should be affixed to the minimum re-order quantity item. For example, attach the Kanban card to the last box of exam gloves size (X) on the shelf. The card should be the appropriate size to visually convey the need. It is important here to note that in order to ensure that special items are identified, a Special Order Kanban Card is created and is filled in by the staff for each of the special order items (*e.g.*, welding gloves, latex-free band-aids). Monitor the Special Order Kanbans to determine if they should be included on the standard Supply Order Form. This will be the case if the special items are used quite frequently and are part of the normal process. A typical card is shown in Table C.3.

TABLE C3
A Typical Re-Order Kanban Card

Supply Re-Order Kanban Card

Item Name:
Maximum Quantity:
Minimum Quantity:
Re-order Quantity:
Supplier Name:
Catalog Page Number:
Return this card to the Kanban envelop

Create standard work: Once the system has been designed and the reordering process has been determined now the team is ready to begin creating the standard work chart. This should be posted in a place where it can be seen and used by the appropriate staff. It can also be used as a training vehicle for the new staff. A typical standard work may be a procedure and or instruction. It can be in a prose form or a process flow diagram.

Conduct the training: The training for the department should be done prior to implementation to ensure the integrity of the system. The training should include the following:

- A brief explanation of the purpose of Kanbans
- Explanation on how the minimum/maximum levels were established and convey appreciation for everyone's input when the Supply Survey was conducted.
- Explanation of how the system will work (distribute process flowcharts).

Explanation of the two types of Kanbans: Supply Re-order and Special Order Kanban; A demonstration at the supply cabinet on how the kanban system will work.

Acknowledgement of the key individuals within the team who contributed to this system.

Communication that this is a work-in-progress trial, and improvement ideas from all staff members will be welcomed.

Implement the kanban system: Training and implementation should occur simultaneously. Once the training has been completed, the Kanban for supplies will be ready to use.

Maintain the standards: After a month or two of usage, review the appropriate budget supply line item, determine the cost savings, congratulate the team, and convey the success of the initiative to management. At all times during the Kanban implementation, take suggestions on how the process can be improved. The responsibility to maintain the system should be rotated among the staff on a regular basis. This will help to ensure, as they take care of the system, that they will have an appreciation for how it works, and therefore, be more supportive of using the system as it was intended to be used. The benefits of implementing a kanban system for supplies are:

- Ensures minimum inventory.
- Creates staff awareness of the cost of supplies.
- Easy tool to implement and train.
- Encourages teamwork.
- Minimizes transactions on ordering supplies.
- Reduces stress.
- Reduces excess inventory waste.
- Allows staff to understand the concepts of flow and supermarkets.
- Reduce costs.
- Eliminates the waste of motion of searching through excess supplies.

FIRST IN, FIRST OUT (FIFO)

FIFO is a work-controlled method to ensure the oldest work upstream (first-in) is the first to be processed downstream (first-out). For example, in central supply areas, some materials are continuously rotated, replacing them with the up-and-coming expiration dates to the front of the shelf. This can also apply to work requests in production, laboratory, tool crib, employee-owned tools, *etc*. The FIFO Lane has the following attributes:

- Located between two processes (supply and demand).
- A maximum number of work units (written orders, lab requests, etc.) are placed in the FIFO lane and are visible.
- Is sequentially loaded and labeled.
- Has a signal system to notify the upstream process when the lane is full.
- Has visual rules and standards posted to ensure FIFO lane integrity.
- Has a process in place for assisting the downstream process when the lane is full and assistance is required.

The team can be creative in establishing the signal method, within the FIFO system, to indicate when the system is full. This could be a raised flag, a light, a pager code, an alert email, or text message to the upstream process. The important point is to ensure the signal established will work effectively. When the signal is released, the upstream worker lends support to the downstream worker until the work is caught up. There is no point in continuing to produce upstream when the downstream process is overloaded. When this happens, it becomes an overproduction waste, which is considered the greatest waste of all. A typical example of this is missing stock between processes and departments.

FIFO lanes help to ensure smooth workflow between processes with little or no interruption. Keep in mind that In-process supermarkets and FIFO Lanes are compromises to true continuous flow. They should be viewed as part of a continuous improvement system, and efforts should be made to find ways to reduce or eliminate them in the quest for continuous flow.

PITCH

Pitch is the time frame that represents the most efficient and practical work (or patient) flow throughout the value stream. It can be a multiple of Takt time. Since Takt time, for many organizations, typically is too small of a unit of time to move the work or information to the next process immediately, pitch is a solution that can be used. Pitch is the optimal flow of work at specific times through the value stream. Pitch is the adjusted Takt time (or multiple of) when the Takt time is too short of a time to realistically move something. Typically, each value stream (or process-to-process timed movement) will have its own pitch. Pitch will:

- Assist to determine the optimal part or workflow.
- Set the frequency for movement of the part or work to the next process.

- Assist in reducing transport and motion waste.
- Allow for immediate attention when interruptions to workflow arise.
- Reduce wait (or queue) times. (Very important note: Each value stream may require a separate pitch.)

Pitch is used to reduce wait time and other wastes that exist within and between processes. For example, a production line was constantly calling to refill the stock bench every time a request from the supervisor had been called in that day. Pitch can be used as a tool in this case to reduce some of the waste that existed in that process. The steps for calculating Pitch are:

Calculate Takt Time

Determine the optimal number of parts or work units to move through the value stream (*i.e.*, number of lab tests to be completed, number of parts to be processed within a specified time period, number of charts to be processed, *etc.*).

Multiply Takt time by the optimal number of work units. Pitch = Takt time (x) optimal number of work units.

For example:

Lab tests requests: 20 tests per day. Time is an 8-hour day (or 480 minutes).

Takt time: Time/Volume = 480 minutes/20 tests = 24 minutes. This means that approximately every half-hour, someone is requesting someone from the production to call in a test for them. Optimal number of tests to be moved = 10 (Ten tests can easily be called in at one time, as agreed to by the staff). This means that:

$$\text{Pitch} = 24\,\text{minutes}\ (\text{Takt time}) \times 10\,\text{tests}\ (\text{optimal number of work units})$$
$$= 240 - \text{minute pitch or 4 hours. This means every 4 hours; 10 tests will be moved to the next process with in the value stream.}$$

Do not confuse pitch with the cycle time. In this example, it only takes approximately 1 minute of cycle time (relaying the test to the lab technician). The total cycle time it takes to complete the 10 tests would be 10 minutes, plus 2 minutes to go to a phone and make the call, for a total cycle time of 12 minutes. If each test were called in separately, then there would be 20 minutes of going to the phone and making the call (each trip 2 minutes × 10 trips). From this example, the waste motion and transport of those other 9 times (or 18 minutes of staff time) would be eliminated. The cycle time would remain the same for each test. (The operator would have been notified of the time that the tests were to be called in.)

Pitch increments must be monitored to ensure they are being met. If an interruption of work arises, a system should be in place to address why the work is behind schedule. There are different ways this can occur:

- The group leader communicates the need for assistance when the pitch cannot be met.

- The employee who cannot meet the next pitch work increment communicates the need for assistance.
- The runner (see next section) communicates the need for assistance.

The communication signal required for assistance can include:

- A pager code or text message to the supervisor or departmental manager.
- An alert email, text message, *etc.* can be sent to all employees for a predefined round-robin type of support.
- A phone call to the supervisor.
- A physical meeting with the departmental leader or supervisor.

Creating a visible pitch board will allow employees to think and work differently about the process that is being improved. A visible pitch board is a bulletin board, whiteboard, *etc.*, displaying the pitch increments and/or associated work that is required to be done for that day or week. The simple fact of having a visual board at a location where the work is being done will be a motivating factor for employees to meet the goal (or pitch increment) that has been determined. This will create a sense of satisfaction each time the work (or pitch increment) is completed. The benefits of a visible pitch board are:

- Begins to remove the work from desks and to make it known.
- Allows managers/supervisors to monitor hourly/daily progress and be aware of situations when work is getting behind.
- Provides a foundation to advance to a more sophisticated scheduling system (*i.e.*, Leveling system) at a later date.

If an employee gets behind (say more than 1 pitch increment) of what is required, there should be a standard procedure to follow (*i.e.*, notify supervisor, request assistance from another staff member, *etc.*). The important point regarding pitch is to group work that can be aligned around a specified value stream, thereby reducing the queue time and other wastes that exist between two processes. Making the pitch increments as a visual cue is a great way to get everyone involved in a work rhythm (*i.e.,* Takt time). It places emphasis on completing a certain quantity of work in an allotted time to meet customer demand and further provides the team with a common purpose of maintaining this "demand." After a specific period of time, usually a month or two of using visible pitch for a value stream, and as other value streams use the visible pitch board tool, then they all can be incorporated into a Heijunka or Leveling system.

WORKLOAD BALANCING

Workload balancing is the optimal distribution of work units throughout the value stream to maintain Takt time or pitch. Also known as employee/staff balancing or line balancing, workload balancing assures that no one worker is doing too much or too little work. Workload balancing begins with analyzing the current state of how work relative to the value stream is allocated and ends with an even and fair distribution of work, ensuring that customer demand is met with a continuous flow mentality. Workload balancing will accomplish the following:

- Determine the number of staff needed for a given demand (or value stream).
- Evenly distribute work units.
- Ensure cycle times for each process are accurate.
- Assist to standardize the process.
- Assist in creating the future state value stream (or process) map.
- Improve productivity.
- Encourage teamwork through cross-training.

The best tool to perform the workload balancing is an Employee Balance Chart. The Employee Balance Chart is a visual display, in the form of a bar chart, that represents the work elements, times, and workers for each process relative to the total value stream cycle time and Takt time (or pitch). The seven steps to workload balancing are:

1. Visually display the list of processes from the current state value stream map: Be very clear about identifying each process, its beginning and end, as well as the process's parameters. The team should have a good understanding of the processes from previously creating the current state value stream map.
2. Obtain individual cycle times for the various process activities: Cycle times should be derived from the current state value stream map and the Document Tagging Worksheet and/or observation data. Revisit these times to ensure accuracy. Members of the team may want to use a stopwatch to time any process activities that may be questionable, evaluating the processes as they check the times. Take time to analyze the processes.
3. Add the individual cycle times to obtain the total cycle time for each process: Each process's total cycle time should match the process times located on the step graph at the bottom of the current state value stream map.
4. Create the Employee Balance Chart of the current state: The Employee Balance Chart is a bar chart identifying each process and staff, along with the various individual cycle times as derived from the current state value stream map. This may be the entire current state value stream map or sections of it where it is known that duties can be shared or consolidated to improve the overall flow. It is recommended that you visually display the chart on an easel with a flip chart so the team can review it as a group and comment. Use Post-it Notes to represent the tasks associated with the processes. Make the Post-it Notes proportional to the time element for each individual task. Draw a horizontal line to represent Takt time.
5. Determine the ideal number of staff for a value stream. When determining the ideal number of workers needed to operate the process or processes for the value stream, keep in mind that staff will most likely have multiple value streams they are responsible for throughout the day. The ideal number of staff is determined by dividing the total process cycle time by the Takt time (or pitch). For example, in the lab, if an average test took 5 minutes (total

cycle time) and the demand (Takt time) was determined to be 5 minutes (12 tests per hour), then the ideal number of staff needed would be 1 person (5 minutes total cycle time/5-minute Takt time). However, it rarely comes out that evenly. Use this to assist in determining staffing levels. However, if the decimal number from the calculation is less than X.5 workers required, balance the value stream to the lesser whole number. Ensure each worker is balanced to Takt time and allocate any excess time to one worker. Utilize the excess time to improve standard work procedures and conduct kaizen activities, attempting to reduce additional wastes within the processes. Once the employee's efforts are no longer needed on the original project, that person can be placed in another continuous improvement capacity or position in the organization. Lean is not about reducing the number of people; it is about eliminating waste. Without this understanding, Lean will never be accepted. If the decimal number from the calculation is equal to or greater than X.5, then balance to the larger whole number.

6. Create the Employee Balance Chart of the future state: Work with team members to move the Post-it Notes around to balance the various work elements with each employee balanced to Takt (or pitch) time, while maintaining the flow of work.
7. Create standard work procedures and train staff: Once a consensus has been obtained on balancing, redistributing, and/or modifying the process activities, update all necessary standards and then train staff to those standards.

STANDARD WORK

Standard work (Standard Operating Procedures) establishes and controls the best way to complete a task without variation from the original intent. These tasks are then executed consistently, without variation from the original intent. Standard work offers a basis for providing consistent levels of healthcare productivity, quality, and safety, while promoting a positive work attitude based on well-documented work standards. Standard work, done properly, reduces all process variation. It is the basis for all continual improvement activities. Creating the standard work procedures comprises of using two main tools: the Standard Work Combination Table and the Standard Work Chart.

The Standard Work Combination Table is the visual representation displaying the sequential flow of all the activities related to a specific process. The Standard Work Combination Table will:

- Document the exact time requirement (*i.e.*, cycle times) for each work element or task in the process.
- Indicate the flow (or sequence) of all the work in the process.
- Display the work design sequence based on Takt time (ideally).
- Demonstrate the time relationship between physical work (patient care, dispersing meds, charting, *etc.*) to the movement of the patient or work (transporting patients, retrieving equipment, checking doctors' orders, *etc.*), queue times, and computer access time.

The Standard Work Combination Table is an important tool for allocating work within the value stream when total cycle times are greater than Takt time. Capturing the motion of the process that is being reviewed is a good method to accurately document the times of each work element. The seven steps for creating the Standard Work Combination Table are:

- Break the process into separate work elements (record motion to obtain accuracy).
- Time each work element from step (1) by observation or EMR/EMS database retrieval.
- Complete the Standard Work Combination Table.
- Review each work element (task). Question whether it should be eliminated or if the time can be improved.
- Gain a consensus on any changes.
- Create a new Standard Work Combination Table. Update the standard work procedures.
- Train everyone on new standard work procedures and audit.

The Standard Work Combination Table is a powerful tool and should be used as the basis for all improvement activities. It does require time to thoroughly complete, but it will be well worth the effort in the long run. It illustrates the sequence of the work performed and it provides a visual training aid for employees. Specifically, it: (a) Displays the work sequence, process layout, and work-in-process in relationship to each other, (b) Displays the worker movement for each activity, task, or operation, and (c) Identifies quality standards, safety concerns, and/or critical opportunities for errors. The process of creating the standard work chart is:

- Draw the area layout or process flow on the chart. Label all items.
- Designate work element locations by number to correspond to items listed in (1).
- Use arrows to show movement of patients, information, staff, and any special instructions or safety concerns.
- Post the chart in the work area and use it as a basis for continual improvement.

It will do little good to improve a process if it is not from an existing standard. If you do not work If you do not work from some type of standard, you will be attempting to improve to a moving target, one of process variation.

Column Chart: Traditional approach to process balancing (**Figures C.1 and C.2**).

YOKOTEN

Yokoten means "best practice sharing" or "taking from one place to another." It encompasses the methods of communicating, documenting, and distributing knowledge horizontally within an organization (peer-to-peer) about what works and what

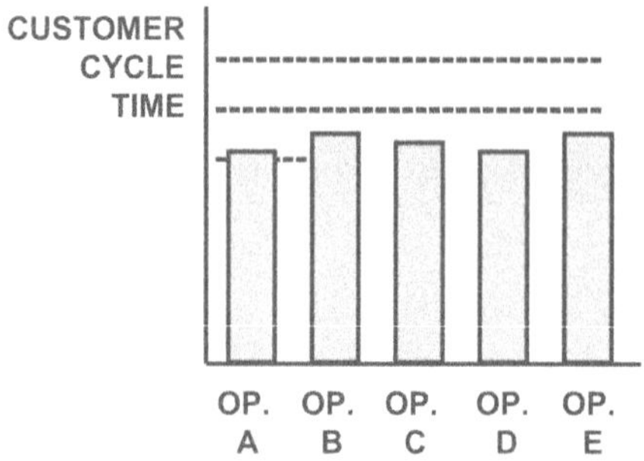

FIGURE C.1 Balancing Processes

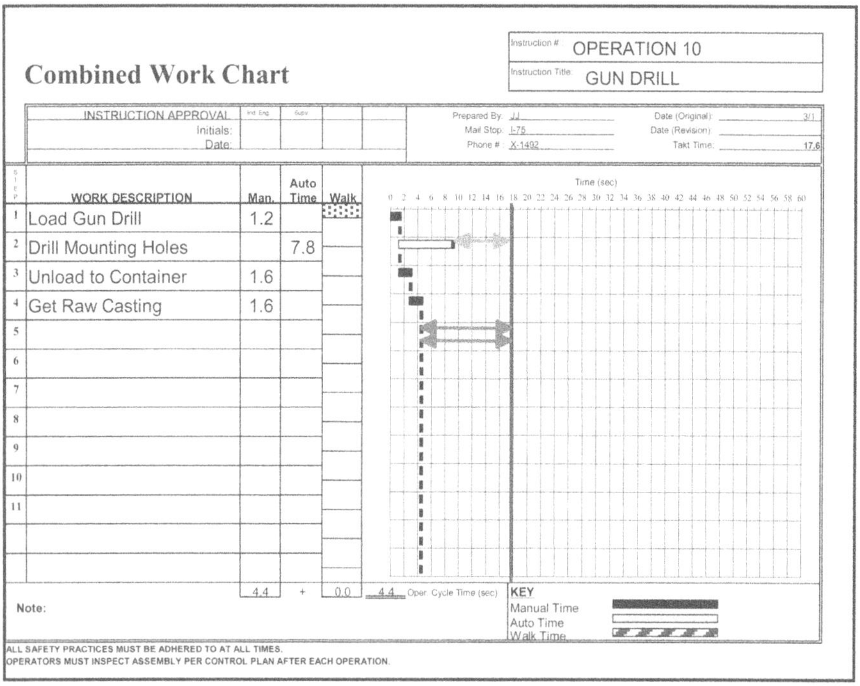

FIGURE C.2 Typical Combination Work Chart

does not work from an improvement project (*i.e.*, PDCA Kaizen Event). *Yokoten* is a form of knowledge management. At its most basic level, Yokoten can be the notebook that a team keeps as a history of the group and problems/solutions encountered. *Yokoten* can be the library of A3 problem reports (Storyboards) that a team or work group maintains for all to access. As a knowledge management device, the Yokoten process ensures information becomes part of the organizational knowledge base. For example, at Toyota, there is an expectation that copying a good idea will be followed by some added "kaizen" to that idea (copy + kaizen = yokoten). *Yokoten* standardizes a solution and shares it. Sharing of standard procedures across an organization is ideal.

Approach of *Yokoten* is as follows:

- Create a standard improvement methodology that problem solving, Lean, or Six Sigma teams will follow. Ensure adequate training is conducted.
- Create standard forms and worksheets to be used. Inform employees where these forms are located on the Local Area Network.
- Assign a certain date and time each month (or quarter) for groups and/or departments to share their PDCA Kaizen Event failures and successes. Allocate 10–15 minutes per group (or representatives) to share their improvement projects.
- Document all completed PDCA Kaizen Events on the network (LAN) or company intranet.
- Ensure each team completes a *Yokoten* Worksheet.

BASIC FLOWCHART (MACRO LEVEL)

Approach to creating this flowchart as follows:

- Define the process boundaries with the beginning and ending points of the process.
- Create no more than six boxes denoting what is occurring between the start and end points.
- Review the flowchart with management to ensure everyone is on the same page with this macro view of the process or workflow.
- Continue with more detailed flowcharting or value stream mapping (if appropriate) and follow the PDCA Kaizen Event methodology.

DEPLOYMENT FLOWCHART

Approach to creating this flowchart is as follows:

- Identify the right people needed to develop the flowchart.
- Define the process boundaries with the beginning and ending points of the process.
- List the major steps (or time element) of the process vertically on the left sheet of paper.
- List the responsible department (or process worker) across the top, each in a separate column.
- List all the steps in their appropriate column and ensure they are connected by arrows.
- Circulate the flowchart to other people within the process for input and/or clarification (if appropriate).
- Identify areas for improvement.
- Create the new flowchart and continue with the PDCA Kaizen Event methodology.

OPPORTUNITY FLOWCHART

Approach to creating this flowchart is as follows:

- Create a process flowchart (micro level).
- Create separate columns on a flip chart or whiteboard. Label one Value-Added and the other one Non Value-Added (or Cost-Added Only).
- List each step from the process flowchart in either column and ensure they are connected by arrows. Expand the steps to show specific areas of concern, if needed.
- Identify areas for improvement.
- Create the new flowchart and continue with the PDCA Kaizen Event methodology.

SPAGHETTI DIAGRAM

Approach to creating this flowchart is as follows:

- Define the beginning and ending of the process.
- Obtain an engineering drawing or create a scale representation of the physical layout of the process.
- List each step of the process on the flowchart (or on a separate piece of paper).
- Label and draw each of the steps (3) sequentially as to how the process flows. Connect each step with an arrow line denoting the flow.
- Gain a consensus if the process is inefficient (*i.e.*, too many touches, too many hand-offs, too much travel, *etc.*).
- Identify areas for improvement.
- Create the new flowchart and continue with the PDCA Kaizen Event methodology.

PROCESS FLOWCHART (MICRO LEVEL)

Approach to creating this flowchart is as follows:

- Identify the right people needed to develop the flowchart. This may require people from outside the PDCA Kaizen Team for their expertise and knowledge.
- Define the process boundaries with the beginning and ending points of the process.
- Define the level of detail required.
- Determine conditions and boundaries for the process flow.
- List all the steps contained within the process flow. The Kaizen Event team may need to walk through the process.
- Circulate the flowchart to other people within the process for input and/or clarification.
- Identify areas for improvement.
- Create the new flowchart and continue with the PDCA Kaizen Event methodology.

PHYSICAL LAYOUT

A Lean physical layout work area is a self-contained, well-ordered space that optimizes the flow of patients and work. Typically, many facilities have areas or departments separated by physical walls that may impede the efficiency of workflow. Eliminating walls may be difficult to achieve, but once people understand some of the efficiencies that can be gained, they become engaged in tackling the challenges. Many different methods can be used, such as the computer and visual controls simulating these walls coming down. It is important to realize that there are delays in many processes due to the physical separations. Figure C.3 shows the comparison between the current physical arrangement with walls and a future Lean state designed without walls. Lean designs without walls remove any and all barriers that impact workflow, improving all types of communication and movement; no design should ever compromise patient safety or confidentiality.

In production, there are at least four possibilities for utilizing physical layout principles. They are:

- Product focused.
- Equipment laid out in process order.
- Straight stream flow.
- Machine closer – see Figure C.4.

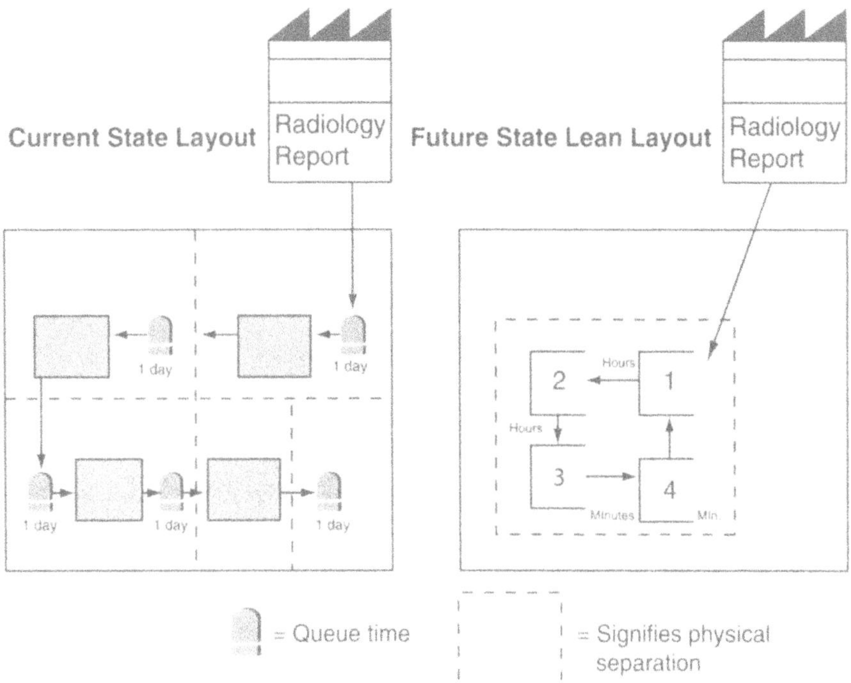

FIGURE C.3 Typical physical layout.

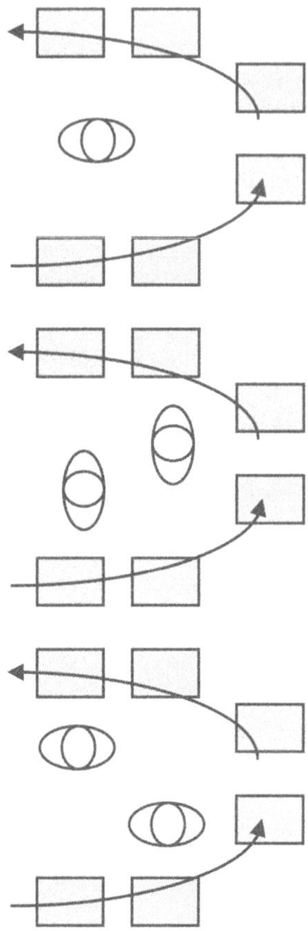

FIGURE C.4 Typical layout configuration.

- Facilitates one-piece flow.
- Reduces non-value-added walking.
- U shape configuration – see Figure C.5.
- Allows volume flexibility with staffing.
- Reduces non-value-added walking.
- Improves communication and visual management.
- Customer focus.
- Modules of capacity matched to major customer sites.
- Allows operation to match customer needs.

The benefits of selecting a physical layout are:

- Process knowledge can be better shared among the staff to allow everyone to be better trained.
- Improved communications will occur between processes.
- Wastes of delay, motion, and transport will be reduced or eliminated.

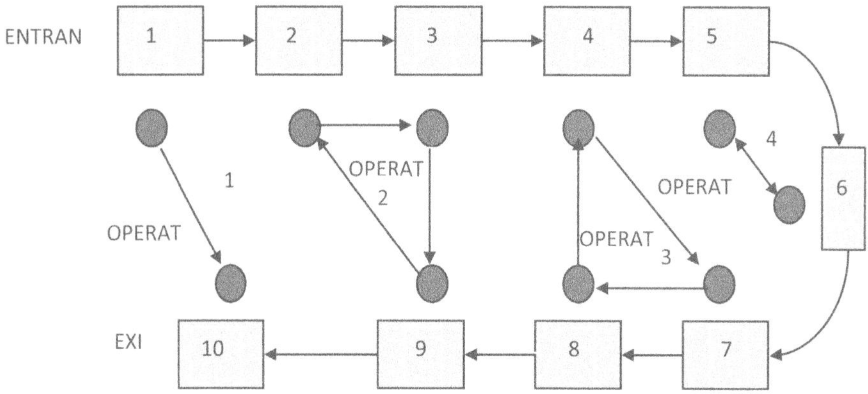

FIGURE C.5 Typical combination layout to take advantage of both equipment and operators.

- Most efficient use of equipment, people, and materials will be ensured.
- Greater job flexibility will occur if someone is absent.
- Less floor space will be required.
- Work throughput will be increased, as well as overall productivity.

The following guidelines can be used when determining a new physical layout.

- Review the current physical layout and associated process tasks to determine which wastes occur due to the current layout in terms of travel, motion, and delays.
- Brainstorm to consolidate where to reduce or eliminate the wastes identified in (1). Processes may need to be modified or standardized and may require additional cross-training.
- Determine if an In-process supermarket or FIFO lane is required.
- Prepare a plan, including result expectations, to implement proposed changes.
- Obtain management approval.
- Implement new layout at a time when minimal disruption to the area would occur. Post any new standards when the new layout has been completed.
- Balance workloads amongst workers and train accordingly.
- Consider new technologies and software enhancements as you continue to improve.

References

Abudi, G. (May 2010). "The five stages of project team development." http://www.pmhut.com/the-five-stages-of-project-team-development.

AIAG (2008a). *Failure Mode and Effect Analysis.* 4th ed. AIAG. Southfield, MI.

AIAG. (2008b). *Advanced Product Quality Planning and Control Plan.* 2nd ed. Chrysler Corporation, Ford Motor Company, General Motors. Automotive Industry Action Group. Southfield, MI.

AIAG. (2009). *Production Part Approval Process (PPAP).* 4th ed. Chrysler Corporation, Ford Motor Company, General Motors. AIAG. Southfield, MI.

Arter, D. (2003). *Quality Audits for Improved Performance.* 3rd ed. Quality Press. Milwaukee, WI.

Arter, D., C. Cianfrani and J. West. (2012). *How to Audit the Process-based QMS.* 2nd ed. Quality Press. Milwaukee, WI.

Barry, Katie, Domb, Ellen and Michael S. Slocum. (2010). "TRIZ – What is TRIZ". The Triz Journal. Real Innovation Network. http://www.triz-journal.com/archives/what_is_triz/. Retrieved 25 July 25, 2014.

Blanchard, K. and E. Parisi-Carew. (2009). *The One Minute Manager Builds High Performing Teams.* William Morrow.

Blanchard-Fields, Fredda (February 2007). "Everyday Problem Solving and Emotion: An Adult Developmental Perspective." *Current Directions in Psychological Science* 16 (1): 26–31.

Bransford, J. and B. Stein. (1993). *The Ideal Problem Solver: A Guide to Improving Thinking, Learning, and Creativity.* 2nd ed. Worth Publishers

Eckes, G. (2001). *Making six sigma last.* John Wiley and Sons. New York, NY.

Fantin, I. (2014). *Applied Problem Solving. Method, Applications, Root Causes, Countermeasures, Poka-Yoke and A3. How to Make Things Happen to Solve Problems.* Createspace, an Amazon company. Milan, Italy.

Ford Motor Company (1998). *Global 8D Participants Guide.* Version 3.0. Ford Technical Education Program, Ford Motor Company. Ford Design Institute Dearborn, MI

Harry, M and R. Schroeder. (2006). *Six Sigma.* Random House. NY.

Hua, Z.; J. Yang, S. Coulibaly and B. Zhang. (2006). "Integration TRIZ with Problem-solving Tools: A Literature Review from 1995 to 2006." *International Journal of Business Innovation and Research* 1 (1-2): 111–128.

IEEE. (2011). *IEEE Guide: Adoption of the Project Management Institute (PMI) Standard A Guide to the Project Management Body of Knowledge (PMBOK Guide).* 4th ed. Retrieved October 30, 2014.

Ishikawa, Kaoru. (1992). *Guide to Quality Control.* Quality Resources. White Plains, NY.

Nikkan Kogyo Shimbun (1988). *Poka-Yoke: Improving Product Quality by Preventing Defects* (p. 111). Productivity Press. Portland, OR.

Offord, Paul (2011). *RPR: A Problem Diagnosis Method for IT Professionals.* Advance Seven Limited. Dunmow, Uk.

Osinga, Frans. (2006). *Science Strategy and War, The Strategic Theory of John Boyd.* Routledge. Abingdon, UK.

Pande, P., R. Neuman and R. Cavanagh. (2000). *The Six Sigma Way.* McGraw-Hill. NY.

Polya, G. (1945). *How to Solve It.* Princeton University Press. Princeton, New Jersey.

Pyzdek, T. (2001). *The Six Sigma Handbook.* McGraw –Hill. NY.

Quality Portal. (May 2013) "Poka Yoke or Mistake Proofing: Overview." *The Quality Portal.* http://thequalityportal.com/pokayoke.htm. Retrieved July 19, 2014.

Robinson, H. (1997). "Using Poka-Yoke Techniques for Early Defect Detection." http://facultyweb.berry.edu/jgrout/pokasoft.html. Retrieved on July 19, 2014.

Sheng, I. L. S., T. Kok-Soo. (2010). "Eco-Efficient Product Design Using theory of Inventive Problem Solving (TRIZ) Principles." *American Journal of Applied Sciences* 7 (6): 852–858.

Shingo, S. (1987). *The Sayings of Shigeo Shingo: Key Strategies for Plant Improvement* (p. 145). Productivity Press. Portland, OR.

Shingo, S. and A. Dillon. (1989). *A Study of the Toyota Production System: From an Industrial Engineering Viewpoint (Produce What Is Needed, When It's Needed).* Routledge, Boca Raton, FL.

Stamatis, D. (1996). *Documenting and Auditing for ISO9000 and QS 900.* Irwin Professional Publishing. Chicago, IL.

Stamatis, D. (2002). *Six Sigma and Beyond: The Implementation process.* Vol VII. St. Lucie Press. Boca Raton, FL.

Stamatis, D. (2002-2003). *Six Sigma and Beyond.* Volumes I-VII. Boca Raton, FL.

Stamatis, D. (2003): *Failure Mode and Effect Analysis: FMEA from Theory to Execution.* 2nd ed. Quality Press. Milwaukee, WI.

Stamatis, D. (2004). *Integrating ISO 9001:2000 with ISO/TS 16949 and AS9100.* Quality Press. Milwaukee, WI.

Stamatis, D. (2004). *Six Sigma Fundamentals.* Productivity Press. NY.

Stamatis, D. (2010). *The OEE Primer: Understanding Overall Equipment Effectiveness, Reliability, and Maintainability.* CRC Press. Boca Raton, FL.

Stamatis, D. (2012). *10 Essentials for High Performance Quality in the 21st Century.* CRC Press. Boca Raton. FL.

Stamatis, D. (2012). *Essential Statistical Concepts for the Quality Professional.* CRC Press. Boca Raton, FL.

Stamatis, D. (2015): *The ASQ Pocket Guide to Failure Mode and Effect Analysis* (FMEA). Quality Press. Milwaukee, WI.

Stamatis, D. (2019). *Advanced product quality planning: The road to success.* CRC Press. Boca Raton, FL.

Stamatis, D. (2021). *Automotive Process Audits: Preparation and Tools.* CRC Press. Boka Raton, FL.

Stamatis, D. (2021a). *Automotive Process Audits: Principles and Practices.* CRC Press. Boka Raton, FL.

Tuckman, B. (1965). "Developmental Sequence in Small Groups." *Psychological Bulletin* 63 (6): 384–99.

Walsh, D. (June 23, 2014). "Automakers, Suppliers BuckleUp." *Crain's Detroit Business.* P. 11–12.

Selected Bibliography

http://www.toolingu.com/beta/definition-350200-5104-maximum-material-condition.html

http://www.tec-ease.com/gdt-terms.php#sthash.kIZbaeuu.JTQQtteQ.dpuf. Retrieved on July 17, 2014.

http://www.ehow.com/list_6911857_layered-audit-procedures.html#ixzz2nrlKipy2

http://www.ehow.com/list_6935021_layered-process-audit-checklist.html#ixzz2nriLnvux

http://www.sustainingedge.com/sustain-with-layered-process-audits/#sthash.VEjrqiJX.dpuf

http://www.iatfglobaloversight.org/docs/Ford_Specifics_for_PPAPJune2013.pdf. Retrieved on September 29, 2014.

http://www.iatfglobaloversight.org/docs/Minimum%20Automotive%20Quality%20Management%20System%20Requirements%20for%20Sub-tier%20suppliers%20-%20AUG%2014.pdf. Retrieved on September 29, 2014.

http://www.barringer1.com/pdf/Chpt1-5th-edition.pdf

http://www.weibull.com/basics/lifedata.htm

Abernethy, R. (2009). *The New Weibull Handbook: Reliability and Statistical Analysis for Predicting Life, Safety, Supportability, Risk, Cost and Warranty Claims.* 5th ed. Robert B. Abernethy. North Palm Beach, FL

AIAG. (2008). *Advanced Product Quality Planning and Control Plan.* 2nd ed. Chrysler Corporation, Ford Motor Company, General Motors. Automotive Industry Action Group. Southfield, MI.

AIAG. (2010). *Measurement System Analysis (MSA).* 4th ed. AIAG. Chrysler Company, Ford Motor Company, and General Motors. Automotive Industry Action Group, Southfield, MI.

Albert, M. (April 2014). "Data Driven Manufacturing." *Modern Machine Shop.* P. 16.

Arendall, C., A. Tiger and K. Westbrook. (July 2014). "Evaluate the Potential of a Process Establishing Process Control." *Quality Progress.* Pp. 28–32.

Bangert, M. (June 2014). "Take These Steps to Six Sigma Success." *Quality.* Pp. 44–45.

Bangert, M. (May 2014). "How to Get the Most Out of Lean." *Quality.* Pp. 40–42.

Bramble, K. (2009). *Geometric Boundaries II, Practical Guide to Interpretation and Application ASME Y14.5-2009.* Engineers Edge. Monroe, GA.

Breyfogle III, F. (1999). *Implementing Six Sigma.* John Wiley and Sons. NY.

Breyfogle III, F. (2003)). *Implementing Six Sigma.* 2nd ed. John Wiley and Sons. NY.

Briggs, S. (September 2012). "(Re)visionary Thinking." *Quality Progress.* Pp. 23–29.

Chaneski, W. (November 2014). "Continuous Improvement Is Vital for Remaining Competitive. Committing to it Across the Organization is Equally Important." *Modern Machine Shop.* Pp. 41–42.

Chaneski, W. (November 2013). "The pulse of a company." *Modern Machine Shop.* Pp. 38–40.

Chatfield, C. and G. Goodhardt. (December 1973). "A Consumer Purchasing Model with Erlang Interpurchase Times." *Journal of the American Statistical Association.* 68. pp.828–835.

Conover, W. (1980). *Practical nonparametric statistics.* Wiley & Sons. NY. J.

Cox, H. (August 2014). "Calibration: Doing Your Own Thing." *Quality.* P. 19.

Cox, H. (July 2014). "Auditor's Guide to Calibration Reports: A Document without Data is Not much of a Report." *Quality.* P. 13

Creel, K. (August 2014). "Working on the Customer's Behalf." *Quirk's Marketing Research Review.* Pp. 78–82.

Cressionnie, L. (July 2014). Revision Runway: SIPOC Implemented in Process for Updating AS9100 standards." *Quality Progress.* Pp. 50–52.

Dailey, K. (2003). *The Lean Manufacturing Pocket Handbook.* D W Publishing. Tampa, FL.

Daimler Chrysler Corporation. (2004). *Process Sign-off (PSO).* 5th ed. Daimler-Chrysler Corporation. Rochester Hills, MI.

DaimlerChrysler. (2010). DaimlerChrysler Customer-Specific Requirements for Use with ISO/TS 16949. http://www.ask.com/web?qsrc=1&o=11517&l=dir&q=DaimlerChrysler+specific+requirements&qo=serpSearchTopBox. Retrieved on September 29, 2014

De Mast, J., W. Schippers, J. Does and E. Van Den Heuvel. (July-August, 2000). "Steps and Strategies in Process Improvement." *Quality and Reliability Engineering International.* Vol 16(4). Pp. 301–311.

Drake JR, P. J. (1999). *Dimensioning and Tolerancing Handbook.* McGraw-Hill. NY.

Dunmire, T. and G. Johnson. (September 2012). "Pointed in the Right Direction." *Quality Progress.* Pp. 15–22.

Evans, J., Foster, S. and K. Linderman. (2014). "A content Analysis of Research in Quality Management and a Proposed Agenda for Future Research." *Quality Management Journal.* Vol. 21(2). Pp. 17–44.

Evers, A. (May 2014). "The Quest for Consistent and Comparable Data." *World Trade 100.* Pp. 28–29.

Ford Motor Company. (2010). *Ford Motor Company Customer-Specific Requirements For Use with PPAP.* 4th ed. Ford. Dearborn, MI.

Galsworth, G. (1997). *Visual Systems: Harnessing the Power of a Visual Workplace.* American Management Association. NY.

General Motors (2010). *GM Customer Specifics - ISO/TS 16949: Including GM Specific Instructions for PPAP.* 4th ed. GM.

Gleason, S. (July 2014). "Bending the Stream of Big Data." *Quirk's Marketing Research Review.* Pp. 50–54.

Gould, L. (May 2014). "FEA: Vehicular Development." *Automotive Design and Production.* Pp. 34–40.

Grachanen, C. (July 2014). "Calculating Uncertainty: Understanding Test Accuracy and Uncertainty Ratios." *Quality Progress.* Pp. 44–45.

Grachanen, C. (May 2014). "Pass or Fail: Understanding calibration content and requirements." *Quality Progress.* Pp. 44–45.

Hayes, R. and S. Wheelwright. (1984). *Restoring Our Competitive edge.* Wiley and Sons. NYJ.

Henzold, G. (2006). *Geometrical Dimensioning and Tolerancing for Design, Manufacturing and Inspection.* 2nd ed. Elsevier. Oxford, UK.

Hinckley, C. (2001). *Make no Mistake.* Productivity Press. Portland, Oregon.

Hinckley, C. M., P. Barkan (1995). "The Role of Variation, Mistakes, and Complexity in Producing Nonconformities." *Journal of Quality Technology.* 27 (3). Pp. 242–249.

Hirano, H. (1995). *5 Pillars of the Visual Workplace: The Sourcebook for 5S Implementation.* Translated by Bruce Talbot. Productivity Press. Portland, Oregon.

Hirano, H. (1996). *5S for Operators: 5 Pillars of the Visual Workplace.* Productivity Press. Portland, Oregon.

Hollnagel, E. (2004). *Barriers and Accident Prevention.* Ashgate Publishing. Aldershot, UK.

Hollows, G. (July 2014). "5 Tips to Maximize Imaging System Performance." *Quality.* P. 1VS.

Howells, R. (June 2014). "Companies Need Real-Time Analytics to Cope with These New Markets." *Worldtrade.* P. 31.

Hutchins, G. (June 2008). "Risk Management – The Future of Quality." *Quality Digest.* Pp. 49–51.

Johnson, L. and M. Deener. (July 2014). "Necessary Measures: Expanded Gage R&R to Detect and Control Measurement System Variation." *Quality Progress.* Pp. 34–38.

Jones, M. (2014). "Identifying Critical Factors That Predict Quality Management Program Success: Data Mining Analysis of Baldrige Award Data." *Quality Management Journal*. Pp. 49–61.

Juran, J. and F. Gryna. (1993). *Quality planning and analysis*. 3rd ed. MacGraw-Hill, Inc. NY.

Lakhal, L. (2014). "The Relationship Between ISO 9000 Certification, TQM Practices and Organizational Performance." *Quality Management Journal*. 21(3). Pp. 38–48.

Littlefield, M. (July 2014). "How closed-loop quality can drive: Near-perfect compliance in A & D." *Quality*. Pp. 19–20.

Liu, S. (May 2014). "Catching Fire: 7 Strategies to Ignite Your Team's Creativity." *Quality Progress*. Pp. 18–24.

Lorden, A., Y. Zhang, S. Lin and M. Cote. (2014). "Measures of success: The role of human factors in lean implementation in healthcare." *Quality Management Journal*. Pp. 26–37.

Marascuilo, L. and J. Levin. (May 1970). "Appropriate Post Hoc Comparisons for Interaction and Nested Hypotheses in Analysis of Variance Designs: The Elimination of Type-IV Errors." *American Educational Research Journal*. 7(3). pp. 397–421.

Marion, R. (2000). *The Edge of Organization: Chaos and Complexity Theories of Formal Social Systems*. Sage Publications. Thousand Oaks, CA.

McCale, M. R. (1999). "A Conceptual Data Model of Datum Systems." *Journal of Research of the National Institute of Standards and Technology*. 104 (4). 349–400.

McClelland, D., J. Atkinson, R. Clark and E. Lowell. (1953). *The Achievement Motive*. Van Nostrand. Princeton, NJ.

Merrill, P. (September 2014). "Embracing Change: Embedding Innovation into Your QMS via ISO 9001:2015." *Quality Progress*. Pp. 44–45.

Mills, C. (1989). *The Quality Audit: A Management Evaluation Tool*. Milwaukee, WI. Quality press. McGraw-Hill Publishing Company. NY.

Montgomery, D. (2009). *Introduction to Statistical Quality Control*. John Wiley. NY.

Nelson, W. (2004). *Applied Life Data Analysis*. Wiley-Blackwell. NY.

Neumann, S. and A. Neumann. (2009). *GeoTol Pro: A Practical Guide to Geometric Tolerancing per ASME Y14.5-2009*. Society of Manufacturing Engineers, Dearborn, MI.

Nikkan, Kogyo (1988). *Shimbun/Factory Magazine Poka-yoke: Improving Product Quality by Preventing Defects*. Productivity Press. Portland, Oregon.

Norman, D. (1989). *The Design of Everyday Things*. Doubleday. NY.

Novak, S. (2006). *The Small Manufacturer's Toolkit*. Auerback Publications. NY.

Papoulis, A. and S. Pillai. (2004). *Probability, Random Variables, and Stochastic Processes*. 4th ed. Wiley and Sons. NY

Parsowith, B. (1996). *Fundamentals to Quality Auditing*. ASQ Quality Press. Milwaukee, WI.

Paterson, M. (1993). *Accelerating Innovation*. Van Nostrand Reinhold. NY.

Peck, R. and J. Devore (2011). *Statistics: The Exploration and Analysis of Data*. Cengage Learning. Stamford, CT.

Phillips, A. (2009). *ISO 9001:2008 Internal Audits Made Easy: Tools, Techniques, and Step-By-Step Guidelines for Successful Internal Audits*. 3rd ed. Quality Press. Milwaukee, WI.

Productivity Press. (2002). *Development Team Kanban for the Shop Floor*. Productivity Press. NY.

QIP/PO Systems. (1991). *Improvement Tools*. QIP/PO Systems. Miamisburg, OH.

Rhinesmith, S. (1993). *A Manager's Guide to Globalization*. Alexandria, Virginia. The American Society for Training and Development. NY. Irwin Professional Publishing.

Rickards, T., & Moger, S.T., (1999). *Handbook for Creative Team Leaders*. Gower.Aldershot, Hants.

Rickards, T., and S. Moger. (2000). "Creative Leadership Processes in Project Team Development: An Alternative to Tuckman's Stage Model." *British Journal of Management*. Part 4. Pp 273–283.

Rocheleau, E. (August 2014). "Calibration Explained." *Quality*. Pp. 31–32.

Rohde, J. (July 2014). "Asking More of the Data." *Quirk's Marketing Research Review*. Pp. 62–67.

Rossett, A. and J. Gautier-Downes. (1991). *A Handbook of Job Aids*. Jossey-Bass/Pfeiffer. San Francisco, CA.

Russell, J. (Ed.). (200). *The Quality Audit Handbook*. 2nd ed. ASQ Quality Press. Milwaukee, WI.

Schaeffler, S. (2013). QAA/S 296001 http://www.schaeffler.com/remotemedien/media/_shared_media/12_suppliers/quality/quality_assurance_agreement__qaa_/history_of_revisions.pdf. Part 1. Advanced Product Quality Planning. Retrieved on August 27, 2014.

Schaeffler, S. (2013). QAA/S29600. http://www.schaeffler.com/remotemedien/media/_shared_media/12_suppliers/quality/quality_assurance_agreement__qaa_/history_of_revisions.pdf. Part 2. Production Process and Product Release Procedure. Retrieved on August 27, 2014.

Schonberger, R. (2014). "Quality Management and Lean: A Symbiotic Relationship." *Quality Management Journal*. 21(3). Pp. 6–10.

Schuetz, G. (August 2014). "Custom Gages to Go: When You Want to Have it Your Way." *Quality*. Pp. 24–30.

Seaman, J. and E. Allen. (July 2014). "Don't Be Discrete: Caution When Categorizing Quantitative Variables During Data Analysis." *Quality Progress*. Pp. 41–43.

Shacklett, M. (August 2014). "Preparing the Supply Chain for the Unexpected." *World Trade*. Pp. 28–31.

Shacklett, M. (June 2014). SCM Spells Success in Emerging Markets." *Worldtrade*. Pp. 28–31.

Shah, D. (September 2014). "Measuring Proficiency: Evaluating Laboratory Measurement Performance." *Quality Progress*. Pp. 46–47.

Sheskin, D. (2004). *Handbook of Parametric and Nonparametric Statistical Procedures*. CRC Press. NY.

Shingo, S. (1986). *Zero Quality Control: Source Inspection and the Poka-Yoke System*. Translated by A. P. Dillion. Productivity Press. Portland, Oregon.

Sobek, D., A. Ward and J. Liker. (March 16, 2000). *Toyota's Principles of Set-Based Concurrent Engineering*. Sloan Management Review. Reprinted from Winter 1999 original. Cambridge, Mass.

Spath, P. (2000). "Reducing Errors Through Work System Improvements." In: P.L. Spath ed. *Error Reduction in Health Care*. AHA Press. Chicago, IL.

Srinivasan, V. (2008). "Standardizing the Specification, Verification, and Exchange of Product Geometry: Research, Status and Trends." *Computer-Aided Design*. 40 (7). Pp. 738–49.

Stamatis, D. (1998). *Advanced Product Quality Planning*. St. Lucie Press. Boca Raton, FL.

Stamatis, D. (2003). *Six Sigma and Beyond: Problem Solving*. Vol. II. St. Lucie Press. Boca Raton. FL.

Stamatis, D. (2004). *Integrating ISO 9001 with ISO/TS 16949 and AS 9000*. Quality Press. Milwaukee, WI.

Stamatis, D. (2012). *10 Essentials for High Performance Quality in the 21st Century*. CRC Press. Boca Raton, FL.

Stamatis, D. (2019). *Advanced Product Quality Planning: The Road to Success*. CRC Press. Boca Raton, FL.

Stewart, D. and S. Melnyk. (2000). "Effective Process Improvement Developing Poka-Yoke Processes." *Production Inventory Management Journal*. Pp. 4148–55.55

Subbiah, S. and D. Bosik. (April 2014). "Instead of Big Data, Try Value Data." *Quirk's Marketing Research Review*. pp. 22–25.
Tandler, B. (October 2012). "Datum Feature: Selection and Use." *Quality*. Pp. 16–17.
Trunick, P. (May 1014). "Compliance Gets Complex." *World Trade 100*. Pp. 20–23.
Tsuda, Y. (1993). "Implications of Fool Proofing in the Manufacturing Process." In: W. Kuo. ed. *Quality Through Engineering Design*. Elsevier. New York.
Tufte, E. (2001). *The Visual Display of Quantitative Information*. Graphics Press. Cheshire, CT.
Ulrich, K. and S. Eppinger. (1995). *Product Design and Development*. 2nd ed. Irwin McGraw-Hill. Boston, MA.
White, A. (2009). *From Comfort Zone to Performance Management*. White and MacLean Publishing. Hoeilaart, Belgium.
Wilson, Bruce A. (2005). *Design Dimensioning and Tolerancing* (p. 275). Goodheart-Wilcox. Tinley Park, IL.
"World watch." *Crain's Detroit Business* Pp. 11–13.
Worthington, J. (June 17, 2014). "How to Manage Change." *Quality Digest*. Electronic version. Pp. 1–2.
Yun, C., Y. Yong, and L. Loh. (1996). *The Quest for Global Quality*. Addison-Wesley. Singapore.
Zchuetz, G. (April 2014). "The Rule of Thumb, Part 1: Guidelines Used to Standardize the Measuring Process Can Provide a Good Basis for Making Gage." *Modern Machine Shop*. Pp.56–58.
Zelinski, P. (September 2014). "Improvement is Not an Option." *Modern Machine Shop*. Pp. 89–99.

QUICK AND VERY USEFUL POCKET BOOK REFERENCES

Baban, B., D. Fertuck, and K. Vermilye. (2009). *The 5S Desktop (PC) Pocket Handbook*. MCS Media.
Bath and Strong. (2002). *Six Sigma: Pocket Guide*. AON Management Consulting. Lexington, MA.
Bens, I. (1997, 1999, 2000). *Facilitation at a Glance*. GOAL/QPC.
Brassard, L. (Ed.). (1985). *The Memory Jogger*. GOAL/QPC.
Brassard, L., and D. Ritter. (ND). *The Memory Jogger II*. GOAL/QPC.
Brassard, M., L. Finn, D. Ginn and D. Ritter. (2002). *The Six Sigma Memory Jogger II*. GOAL/QPC.
Hadfield, D., S. Kozlowski, and T. Speri. (2009). *The Lean Healthcare Pocket Guide*. MCS Media.
Martin, P. and K. Tate. (1997). P*roject Management Memory Jogger*. GOAL/QPC. Methuen, MA.
Munro, R. (2002). *Six Sigma for the Shop Floor: A Pocket Guide*. ASQ Press. Milwaukee, WI.
Munro, R. (2003). *Six Sigma for the Office: A Pocket Guide*. ASQ Press. Milwaukee, WI.
Munro, R. (2009). *Lean Six Sigma for the Healthcare Practice*. ASQ Press. Milwaukee, WI.
The Simply. Lean: Pocket guide. MCS Media.
Stamatis, D. (2015). *The ASQ Pocket Guide to FMEA*. ASQ Quality Press. Milwaukee, WI.

Index

Note: *Italic* and **Bold** page numbers refer to figures and tables, respectively.

D

E

F

G

H

I

J

K

L

M

For Product Safety Concerns and Information please contact our EU representative GPSR@taylorandfrancis.com
Taylor & Francis Verlag GmbH, Kaufingerstraße 24, 80331 München, Germany

www.ingramcontent.com/pod-product-compliance
Ingram Content Group UK Ltd.
Pitfield, Milton Keynes, MK11 3LW, UK
UKHW021301190926
475701UK00009B/268
9781041095316